T0269767

Botany of the Kitchen Garden

Botany of the Kitchen Garden

The science and horticulture of our favourite crops

Hélèna Dove
With illustrations by Martin Cutbill

Kew Publishing
Royal Botanic Gardens, Kew

Tomato
'Gardeners
Delight'

Contents

Plant types

Introduction

We have been growing fruits and vegetables in domesticated situations for millennia, from large-scale agriculture to windowsill pots. Most people have grown their own at some point. Food growing is an important part of cultures and lifestyles, often bringing people together to grow and share a meal.

Over the years, there has been a large variety of literature produced on how to grow crops. The advice changes from era to era, with changes in attitudes and accessibility to new crops and cultivars. Many wartime manuals on growing food in the UK speak about using nicotine as a homemade pesticide, whereas more modern texts focus on growing organically with no chemicals at all. Newer information covers crops that previously may not have been suitable for a country, but with changing climates and crop breeding can now produce a harvest, such as the sweet potato in the UK. Information covers large-scale farming, focusing on monocultures and how to gather the largest output suitable for a supermarket, but also windowsill and container growing in the smallest spaces, with dwarf cultivars and even microgreens. Amid this abundance of information about horticultural practice, there are dotted around gems of wisdom about the fruits and vegetables themselves.

Botany of the Kitchen Garden pulls together this information, focusing on the crops: which part of the plant we harvest, how are plants adapted and how this affects the grower. Some horticultural practices are also investigated with a scientific slant, giving a deeper understanding to an afternoon spent at the allotment.

All of the crops we eat and grow have wild ancestors, some of which can still be found today and some, sadly, which can't. Humans have taken these wild plants and domesticated them through cultivation and breeding. This process has enhanced qualities that we find appealing. In many cases, it has been to enlarge the target, edible part of the crop – from the swollen terminal bud of the cabbage, to the root of the carrot, which can now reach over a metre in length, and types such as oxhearts that have very wide shoulders in comparison to the spindly root of the wild carrot. When grown for their fruits, sweetness levels have often been increased, and the fleshy parts swollen through breeding. In some, more recent cases, the public demand for uniformity has led to consistency of colour, size and shape, and in some fruits, seedless varieties have been extremely popular.

Plant breeders have also made harvesting easier. Uniformity of ripening on vegetables such as Brussels sprouts, which generally slowly mature from the bottom up, so that the entire plant can be harvested at once, makes life much easier for the now mechanised process. Plants that have thorny pest defences such as blackcurrants have been bred to have thornless varieties to make the picking process less painful. Tomatoes have been bred with thicker skins, which bruise less easily when being transported to market.

Domestication and breeding have led to some exciting developments and interesting cultivars, but they have also meant that many plants lack their natural defences again pests, diseases and a changing climate. This has led scientists to go back to the plant's wild relatives, thriving in their natural habitats. The hope is that breeding some of this resilience back into the crops grown will lead to healthier ecosystems where growers rely less on chemicals for strong, healthy crops. But going back to the wild plant also helps to understand the crops in other respects, and these can help the home grower decipher why their fruits and vegetables behave in certain ways, especially examining their origins and botany.

By examining the botany of both wild and domesticated plants, we can understand behaviours, both natural and in response to external factors such as too much or too little heat. Knowing where

a plant originated from in the wild can explain certain adaptations that need to be considered when growing, and also how to provide the best conditions – or even if the plant will grow at all.

Knowing that in the wild, rhubarb thrives alongside water courses means in the garden it needs plenty of moisture in the ground to thrive. Tomatoes come from warm climates that do not experience frost, so they need heat to break dormancy and grow, but they will not live through a frozen winter. Crops that originate in humid environments, such as the pineapple, will never prosper outdoors in a British kitchen garden, where humidity is never high, and nor are the summer temperatures. To grow pineapples in these conditions, glasshouses and protected cropping come into play. Plants that come from coastal regions, such as members of the brassica family, experience fairly constant temperatures in the wild and do not undergo the more extreme highs and lows of inland kitchen gardens. To get around this, growers can grow these crops at times when temperatures are more steady.

The botany of a plant tells us more about its adaptations and how to recognise it. But these features can again play into helping growers produce the best crop. Botany covers many different aspects of plants from a scientific point of view. It examines plant biology, the fundamentals of how the plants work. It can include how they produce energy, via several different adaptations of the process of photosynthesis, looking at what happens within a cell, and how plants can adapt to refine the process with physical adaptations such as larger leaf areas.

Botany also includes genetics and classification, how plants relate to each other and how their genes affect them. Understanding that potatoes and tomatoes are in the same family and are genetically similar means growers are aware that if their potatoes get the disease blight, then their tomatoes may also succumb. When faced with a new crop, a grower can call on their experience of another plant in the same family or species, and this can aid them gaining success as the plants may have similar needs when growing. Genetics is becoming more and more important, especially during the breeding process. Scientists are able to isolate genes for factors such as sweetness or colour, maybe even disease resistance, and this information can aid the breeding process.

Botany also includes the structure and physiology of plants. As a grower, understanding whether the crop you are trying to grow is a storage organ, a fruit or a leaf can help guide cultivation for maximum output. Knowing that courgettes are fruits means a grower can feed the plant once fruiting occurs for maximum output. Or understanding that a beetroot is not a true root crop, but formed of the upper part of the root and lower part of the seedling stem, helps a grower to know that it will transplant happily, unlike a true root crop such as a carrot. As the carrot stores reserves in its tap root, any damage to the tap root during transplanting can lead to a deformed, forked crop.

The life cycle of a plant will also inform the cultivating and harvesting of a crop. The aim of any plant is to reproduce, and

often it is this that the grower is manipulating and fighting against. Many plants are triggered to flower by day length, known as photoperiodism. Florence fennel will flower when the day length is 14 hours, generally the middle of summer, as in the wild these are the perfect conditions for spreading seed. Knowing this, the grower either needs to breed this trait out, or not attempt to grow Florence fennel in the middle of the summer months. Knowing that runner beans are perennial in the wild, even though they are grown as annuals in most kitchen gardens, can explain why any roots left underground in milder climates have a tendency to shoot in spring, disrupting the neat rows of young vegetables that have been planted this year. As a biennial, a parsnip will not produce seed until

its second year, so for those wanting to save seed, the odd root must be left in the ground until the next season. Hardneck garlic cloves must undergo cold to produce a flower stalk, which in turn produces cloves. In the wild, if the garlic experiences cold, and then warmth, the plant knows it is the temperatures of spring and not autumn it is about to encounter. All this information can be gleaned from researching the botany of fruit and vegetable crops, and can help improve the output of the kitchen garden.

Vegetable gardeners are often keen on saving seeds to sow again the following season, or to gift to other growers. Understanding how a crop produces its offspring can ensure success and avoid disappointment. Knowledge about the botany of a crop can reveal whether a plant bears both male and female flowers, or if they are on separate plants. It will also identify if a species can self-pollinate or whether it needs to cross-pollinate, as is the case with apples. If plants cross-pollinate, it is unlikely the seeds will produce identical clones of their parent, so growers can decide to control pollination and isolate the flowers to ensure healthy offspring with the desired qualities. Knowledge about how plants are pollinated, be it by insects or wind, suggests whether plants must be planted relatively closely.

The fascinating world of edible plants is a large and wondrous one, and the crops within this book are simply some of the most commonly grown. Beginning to understand the language of botany and how this relates to a plant, the plot and, eventually, the plate will give readers and growers the tools to delve further into their favourite foods. This book will hopefully inform future growing, and give a new appreciation of the plants that are sown, cultivated and devoured.

Plant nomenclature and taxonomy

In most vegetable gardening texts, vegetables are referred to by their common names – carrots, apples and tomatoes, for example. But these plants also have another, more official name, their botanical Latin name. This name describes a plant's position within the biological hierarchy of living entities. The naming of plants is known as nomenclature, but deciding where plants fit is decided with botanical taxonomy. It is a system that is open to change as science evolves and taxonomists improve their understanding of the living world.

Much taxonomy was originally based on physical observations, such as the number of petals or arrangement of leaves. There is now a new system that has been recently developed known as the APG or Angiosperm Phylogeny Group system, which uses DNA and molecular technology to classify plants. This has led to some name changes and movement within the taxonomic ranks, so in older texts plants may have different names, known as synonyms.

All plants can be classified by a minimum of seven taxonomic ranks, which group them by characteristics.

First, kingdom is the highest rank and includes animals, plants and fungi. Within the kitchen garden, most plants fall into the

plant kingdom; if growing edible mushrooms would fall into the kingdom of fungi.

Below kingdom is phylum, which is sometimes known as division, of which there are 12 categories. Most crops fall into the same one because they have flowers, roots and stems.

Below this is class, and then order. More recently, there has been a change to taxonomy where the divisions are renamed as clades, which is a developing line of classification. Mainly these higher classes are not referred to in general texts, but the lower three are, starting with family, then genus and finally species.

The family groups together plants by common characteristics and will always end in -aceae in texts. Understanding what family a plant belongs to may help to understand how it will behave in the kitchen garden. For example, the tomato, *Solanum lycopersicum*, is found in the family Solanaceae, a family that predominantly likes heat. Citrus fruits, which are in the Rutaceae, are likely to be evergreen shrubs and have oil-producing glandular leaves.

The next two taxa are genus and species, and together these create the binomial Latin names of plants, which are most commonly cited in texts.

The genus is the higher classification, and plants grouped within a genus will have a significant number of shared characteristics, such as the same flower parts or hollow stems.

Within the genus are the species. Members of a species can interbreed and produce fertile offspring, which in turn can produce young themselves. Information such as this is crucial in vegetable breeding and seed saving. At a time when many crops growing in the fields have been bred to the point they have no resistance to a changing climate, understanding that resistant traits from wild species can be bred back into them could be vitally important in feeding the world.

There are more botanical classifications below species, known as subspecies, variety and form, but not all plants have these.

A subspecies, indicated by the term subsp., is created when a species has grown in two distinct geographical locations, which has, over time, created different characteristics in the populations, but not enough to form a new species. Quite often the subspecies are left off when referring to the cultivation of fruits and vegetables; for example, squash, *Cucurbita pepo*, has several subspecies, including *C. pepo* subsp. *pepo*, which includes courgettes and *C. pepo* subsp. *texana*, which includes crookneck squash, but fundamentally both are cultivated in the same way.

Variety, var., is very similar to a subspecies, but the differences are within a population that are in the same geographical location. Varieties tend to hybridise freely, which often causes these small differences.

A form, f., has very slight differences within a population, such as a slight difference in flower colour, often caused by a genetic mutation. These lower classifications are useful in the kitchen garden as we know to expect slight variations in our plants, and that a slight variegation in a leaf, for example, may just be a natural mutation, and not necessarily a disease.

Scientists are constantly botanising and researching plants, and sometimes the names and classifications of plants may change with expanding knowledge. Occasionally, the lower taxonomic ranks are ruled to be botanically invalid, but within a garden setting, these differences may be important. These ranks are then classified as groups.

One of the most infamous groups in the kitchen garden lie within the brassica family. Cabbages, cauliflowers, sprouts, kale and many others share the same binomial Latin name of *Brassica oleracea*. Historically, the differences in these plants has been

recognised as varieties, but more recently they have been classed as groups, so now a cabbage will find itself referred to as *B. oleracea* Capitata Group.

As most crops have been bred in cultivation to enhance certain characteristics, such as sweetness, size and colour, they find themselves with an extra taxon, known as the cultivar. The cultivar will also often appear after a common name in seed catalogues, so *Zea mays* 'Swift' may be written as Sweetcorn 'Swift'. In the kitchen garden, cultivar choice is incredibly important and one of the most widely discussed topics. Everyone has their favourites.

One last botanical designation of plants is the hybrid. These can arise via cross-pollination, with the offspring having distinct characteristics from the parents, and can be formed naturally or with a little help from humans. The majority of hybrids in the kitchen garden are formed by crossing two different species, known as interspecific breeding, and the resulting plant has an 'x' in its name. One of the commonest examples is the apple, *Malus* x *domestica*, which is descended from many wild plants that hybridised naturally to produce the trees grown today.

Although it is unlikely that botanical Latin will start to be the main talking point at the allotment, it is a useful tool. Common names have the downside that they can be quite colloquial – for example, a *Cucurbita pepo* is known as a courgette to some and a zucchini to others, or the common name of turnip refers to both a *Brassica rapa* and *B. napus* depending on who you are speaking to.

Plant botany

Although plants differ wildly, there are some features that are fairly universal. The descriptions of the plant parts in this section refer to those most commonly found and do not delve into the modifications in individual species. For example, roots are referred to as growing underground, but in epiphytic or air plants they are aerial parts, although still performing the same basic function of taking up water.

Below ground are the roots, whose main function is to absorb water and minerals, but also anchor the plant into the ground, plus store food in their cells. There are two main types of root system, namely a fibrous root system, which consists of lots of small roots and tends to grow in the top parts of the ground, while other plants have a tap root, or primary root, which grows directly down, and coming off the sides of this are the lateral roots. There are also root systems that are a combination of the two types. Root hairs appear on the roots, whose job is to increase the surface area of the root and allow more water uptake.

The aerial parts of the plant generally revolve around a central stem, which provides support for the plant. In woody plants this is known as the trunk, which has branches coming off it. Along the stem are the nodes, which are areas of growth where axillary and leaf buds are found. The leaf buds form the leaves, and the axillary buds grow in axils, the place where the leaf joins the stem, and can form vegetative shoots and reproductive shoots. The leaves' main function is as the site of photosynthesis, which creates energy for the plant from the sun's light. The leaf itself is formed of a petiole or leaf stalk and the blade, which is the surface of the main leaf. At the top of the stem is the apical bud, which is the primary growth point, allowing the plant to grow upwards.

Another feature of most plants is their flowers and fruits. The flowers contain and protect the reproductive organs, and are also designed to attract pollinators. A fertilised flower will turn into a fruit, which contains within it the seeds for the next generation of that plant. The function of the fruit is to protect the seeds until they are able to germinate, but in some cases also to attract dispersers such as birds, mammals and insects. Plants that produce flowers and fruits are called angiosperms, but there is a large group of plants called gymnosperms that do no produce flowers and their seeds are not contained within a fruit. Gymnosperms do not feature heavily in most modern kitchen gardens.

Botany of a flower

The flower contains the reproductive parts of the plant, and can be either complete (contain both male and female organs), or incomplete (just male or female). The eye is mainly drawn to the outer parts of the flower, the petals, which are not reproductive. The stalk of the flower is known as the pedicel, and the top of the stalk, which is flattened and contains the flower parts, is referred to as the receptacle or hypanthium.

The outer parts of the flower are collectively called the perianth, which comprises the calyx and corolla. The outer whorl is known as the calyx and is formed from sepals, whose main job is to protect the flower; in many cases they are only really visible when the flower is in bud form, where they guard the outer layer. The corolla is the next whorl inwards and consists of the petals, which are brightly coloured to attract pollinators. The perianth on some flowers consists of only one whorl, formed of tepals, which are modified leaves that change colour during the life of the flower.

The female flower part is known as the pistil or carpel, and is formed of the stigma, style and ovary. The stigma sits at the top and has a sticky surface to collect the pollen grains from the male. The style is a long tube, which gives the stigma height to ensure it is in a place to easily catch the pollen. At the base of the style is the ovary, which is eventually fertilised and forms the embryos, seeds and fruit for the next generation.

The male organ is known as the stamen and consists of the anther, where the pollen forms and disperses from, and the filament, which raises the anther high enough to effectively spread the pollen.

Plant

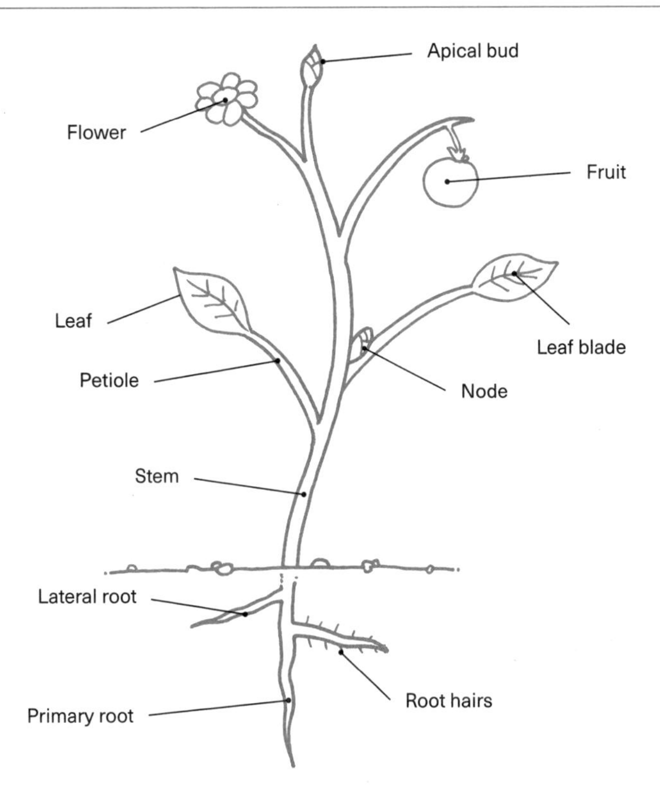

Flower

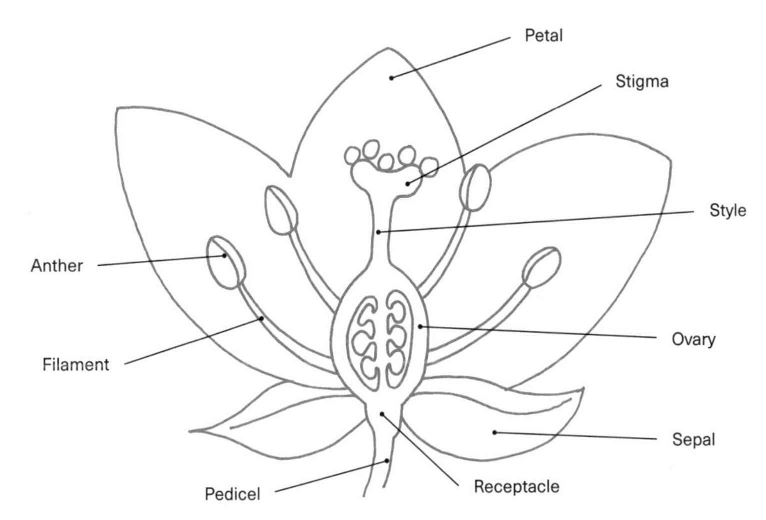

Plant types

Abelmoschus esculentus

Malvaceae

Okra, Bhindi, Gumbo, Ladies' fingers

The fruits of *Abelmoschus esculentus* have many names, from okra to gumbo, and are eaten around the world. This plant is a cultigen, meaning it is only known in cultivation with no traceable wild ancestor. Despite this, it is thought to have been in cultivation for thousands of years, starting with the Egyptians in 2000 BCE – okra can still be found growing along the banks of the river Nile. It is a tender, herbaceous plant that likes a lot of moisture and can grow up to 2 metres (6.5 ft) tall in optimum conditions, with temperatures of 21–30°C (70–90°F).

The plant itself has leaves that are edible when young, and start off in an oval shape, maturing to a lobed leaf with notched margins. The ephemeral flowers open and close remarkably quickly, but have large cream petals with a stunning red centre. The petals are imbricate, meaning they overlap, and so it can be seen why okra was once thought to be a hibiscus, with its synonym being *Hibiscus esculentus*. The flowers are produced singly in the lead axils and are hermaphrodite, producing both male and female organs, and self-pollinating.

Once pollination occurs, the fruits are quickly produced, and best harvested in the first two days, otherwise they turn woody and inedible. The young fruits are often referred to as pods, but botanically speaking are loculicidal capsules which split lengthways, consisting of five chambers, which contain the seeds and give a pentagonal-shaped fruit. As well as the clearly marked sides, the capsules are also slightly hairy and can be green or burgundy.

On the end of the okra is the cap, which develops from the base of the flowers where the petals and stamen were attached, and occasionally bracteoles can be seen on the cap margin. Within a few days, the capsule will become woody, drying to split along the walls of the chambers to release the seeds, making it a dehiscent fruit. The dried seeds can be used as dried beans, and have been known to be roasted and used as a coffee substitute.

One of the most striking things about immature capsules that are harvested is their slimy texture. This slime is a water-soluble acidic polysaccharide, known as mucilage. Although it is water soluble, it isn't easy to wash away, but is perfectly edible and is used as a thickening agent in stews. It is not known why *Abelmoschus esculentus* produces this mucilage, as it seemingly does not deter predators and pests, and does not aid growth, unlike the mucilage produced by root caps which lubricates the roots' journey through the soil.

Okra

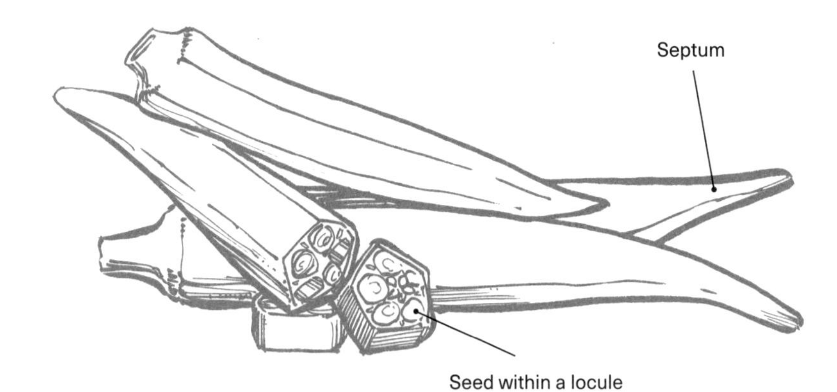

Actinidia chinensis var. *deliciosa*

Actinidiaceae

Kiwi fruit, Chinese gooseberry

A native of south-west China, *Actinidia chinensis* var. *deliciosa* achieved its most common name of kiwi fruit for its popularity in New Zealand. The kiwi is a liana, or woody climber, with flexible stems, and grows up to 10 m (32 ft). The vine itself is hardy to –18°C (–1°F), although the flowers, fruit and new growth are susceptible to frost damage. Although in the wild it grows on the banks of rivers, it has many adaptations to drought, including waxy uppers on its cordate ('heart-shaped') leaves to reduce transpiration, and if its roots become too dry, it will close leaf stomata, again to reduce water loss.

The flowers are formed once the stems reach their second year. Many woody fruit plants begin floral induction, the process where the shoot apical meristem becomes able to develop flowers, in the previous summer, and then produce flowers in the following spring. The kiwi starts floral induction in the spring that the flowers are produced, meaning if the plant experiences any stress in this time period, flower production can be impeded.

The flowers themselves are cupuliform, or cup-shaped, with five white petals. The vines are dioecious, with separate plants being either male or female, although there are now some cultivars, such as 'Jenny', that are hermaphrodite and self-fertile. The male flowers are small in comparison to the females. The white flowers are not visually attractive to their main insect pollinators, but instead are strongly fragrant. Although predominantly entomophilous, or pollinated by insects, pollen can be carried by the wind from one flower to the other, so they are also described as anemophilous. Insect pollination tends to be more successful, as more pollen is brought to the flower, creating more seeds. A fruit can have up to 1,500 seeds, and needs at least 800 to give a good fruit size.

The fruit is an oval berry with green flesh and epidermis but it appears brown because of the fuzz on its surface. Some kiwis can be golden and even red. The eventual flesh colour is related to the hormone cytokinin. All kiwis start with green flesh, where chlorophyll is the dominant pigment and useful for photosynthesis, but once cytokinin production stops, the chlorophyll starts to degrade, revealing other pigments and giving a varying flesh colour depending on these. In green-fleshed kiwis, it has been shown that they produce different and high concentrations of cytokinin, which results in chlorophyll degrading less and giving a green flesh.

Within the fruit, the flesh is derived from a single ovary and surrounds a central columella, which is then surrounded by two layers of pips, which turn black once the fruit is ripe. The fruit produces the enzyme protease, whose job it is to break down proteins, which is why kiwis can leave a tingly sensation on the lips and hands when consumed.

On the vine, the kiwi is a brown colour, which for a fruit that wants to encourage herbivores to eat it and spread its seed seems an unusual choice as it doesn't stand out and compete with nearby edibles. But this fruit is seemingly designed to attract primates, with one of its older common names being 'Míhóutáo' or macaque peach in Mandarin. Primates care more for size, and prefer fruits in the brown or orange colour range. The brown-coloured fuzz is not actually hairs, but multi-cellular trichomes that originate in the cell walls of the epidermis. These non-glandular hairs vary in length and help the fruits to combat the elements. They deflect sunlight, and also stop high winds from dehydrating the fruits.

Actinidia arguta

Kiwi berry, Siberian gooseberry

This relative of *A. chinensis* var. *deliciosa*, the Kiwi berry, has a much smaller fruit, but fuzz-free, making it an attractive proposition. The

Kiwi

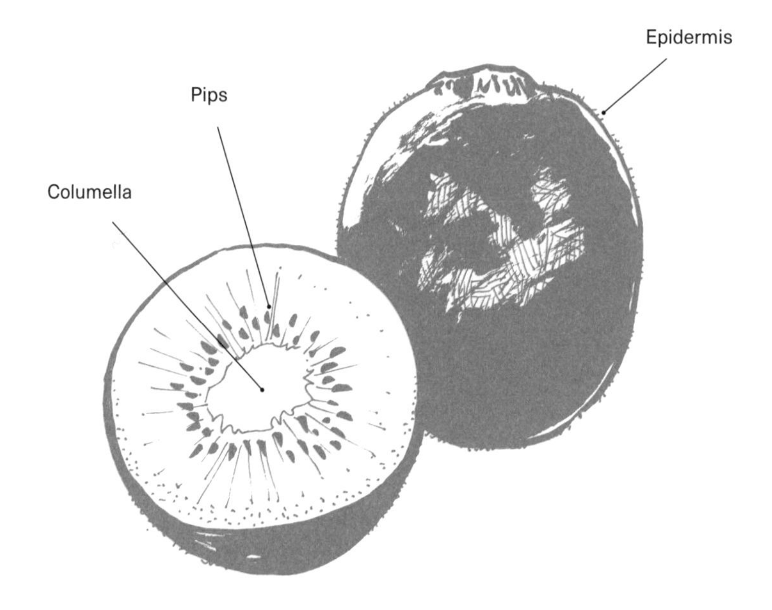

Leek

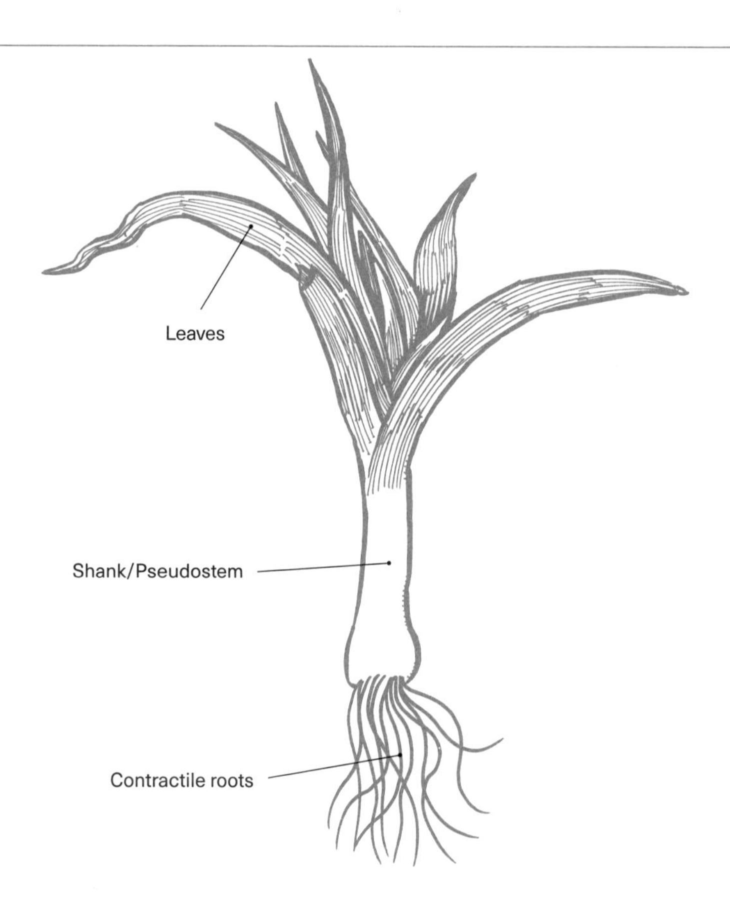

small fruits are green-fleshed, but can also be blushed red. This liana is much more hardy than its cousin, down to –35˚C (–31˚F), but is similar in that it is dioecious, though it has some hermaphrodite cultivars.

Allium ampeloprasum

Amaryllidaceae

Leek

Allium ampeloprasum is commonly known as the leek, and for much of its history has been referred to as *A. porrum*, but is now settled in terms of its Latin name. A member of the Allium genus, which is notable for its strong 'onion' flavour, the leek has a slightly milder taste. Native to the Mediterranean and Central Asia, it is grown and harvested for the white shaft at its base, which is referred to as a pseudostem, as it is in fact a group of tightly overlapping leaves and not a stem. Many plants in the genus Allium build reserves in these lower storage leaves and create a bulb, which is what they are grown for, but the leek only creates a small bulb and is harvested before this visibly occurs.

The bottom of the leek is not naturally white, but instead is blanched by gardeners, either by planting in a trench and backfilling or mounding up the plants. This process removes light from the pseudostem, which stops it producing chlorophyll and makes its sweeter and more tender. The top, green leaves are also edible, but much tougher and are often discarded. Leeks are monocots and have many obvious traits, including slim leaves with parallel transport systems.

Underground, the leek produces contractile roots, which help pull it down to a secure depth. Contractile roots appear slightly wrinkled when the leek is harvested. They start life with a smooth surface but with time the root will start to shrink, while the root tip stays where it is, shortening the root length and pulling the plant down. This process is a response to the amount of light the roots detect, specifically in the blue spectrum. Once the roots reach a depth where they are receiving the desired levels of blue light they stop shrinking.

The leek is harvested as an annual in its first year of growth, often in the winter as it is extremely hardy, but if allowed to develop for a second year, it will develop a flower head that produces small black seeds, which is how leeks are typically propagated. If left in the ground after the flower is formed, the leek will actually grow as a perennial as it produces a bulb underground, which splits and produces several pseudostems for harvesting, leaving some individuals to perennialise.

Allium ampeloprasum* var. *Babingtonii
Babington leek
This coastal variety of leek is truly perennial, and will re-grow when cut at the basal plate. It will form a bulb underground, which, like the straight species leek, will split and form new plants. It is also a topsetting allium, producing bulbils in its flower stalk, which will drop and form new plants.

Allium ampeloprasum* var. *ampeloprasum
Elephant garlic
Elephant garlic is in actual fact a leek, but one that is harvested for it massive bulb, which contains similarly massive cloves. Milder than true garlic, it is a popular garden curiosity.

Allium cepa

Amaryllidaceae
Onion, Spring onion, Scallion

Allium cepa is a popular kitchen garden crop used as the basis for many meals, and best known by its common name of the onion. Although it is not found growing wild, it is believed to have originated in the Mediterranean and south-west China. It is a cool-season plant, and runs to seed or bolts if the temperatures get too high. The onion is a biennial, herbaceous monocot, producing a flower in its second year, and storing energy in its first.

Although all parts of the onion are edible, it is the bulbous storage organ that is most commonly sold in shops. A bulb is the area of *Allium cepa* where the energy it produces during photosynthesis is stored as sugars, unlike many storage organs that retain the energy in more complex carbohydrate forms. The layers of the onion are formed of modified leaves, which can be easily seen once harvested and cut open. The bulb sits upon a basal plate, a modified, flattened stem that has roots on the underneath and needs cutting away when preparing the crop for eating.

The leaves of the onion have two distinct parts separated by a membranous ligule. The lower, non-photosynthetic parts of the leaf form the storage area, while the upper, green, photosynthetic part is hollow and can be cylindrical or flattened. These leaves are also edible, but much less commonly consumed

Onion

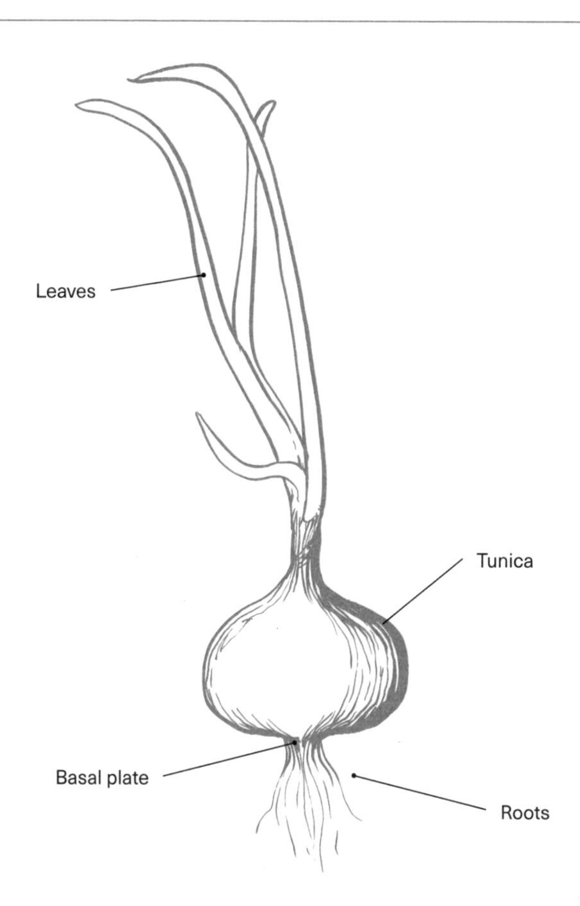

when mature, although they can be cropped when young and the bulb hasn't formed yet. When the onion is harvested young, it is known as a spring onion or scallion.

The onion is a tunicate bulb, which means it is surrounded by a dry fibrous layer, known as the tunic or tunica. This layer is an adapted leaf and defends the bulb. The storage leaves of the onion is formed in the centre of the bulb from nodes on the basal plate. Onions are triggered to start bulbing and storing food by day length, a trait known as photoperiodism. Not all onions are triggered by the same photoperiod, and there are short-day onions, which start bulbing when day length is 10–12 hours, intermediate cultivars, which need 12–14 hours, and long-day onions, which need 14–16 hours. This often answers the question why certain onions do well in comparison to others, and can change drastically between the north and south of a country.

Once the conditions are right, in the second year, the bulb will send up an inflorescence, an umbel with several small flowers. These are pollinated by insects, particularly flies, and will produce seeds. Seed is one way of propagating onions as a gardener, but there are also sets. Onion sets are started by seed in a nursery during the season before they are to be grown to harvest. The seed is thickly sown and produces lots of small bulbs, which are harvested and stored at an immature stage. The next year, they are sold as sets, which gardeners plant in the spring and will bulb up by the summer.

Onions, and many other plants in the Allium genus, are easily recognised by their strong taste. As is common in plants, strong tastes are produced as a pest defence, and in the case of the onion, against herbivores in particular. The plants' overwhelmingly potent taste evolved to deter herbivores from eating them. Ironically, humans grow onions, just for this taste.

The onion flavour starts life as a chemical called cysteine sulphoxide, which is stored in the cytoplasm of the cells. Once these cells are damaged by a herbivore, the enzyme alliinase, which is stored in the cells' vacuoles, meets the cysteine sulphoxide and transforms it to sulfenic acid, giving that instantly recognisable flavour. In addition to these, another enzyme is released called lachrymatory factor synthase, which converts the acid into a volatile gas. This volatile gas stimulates the lachrymal glands in the eye, and causes humans to produce tears. The gene that encodes the production of lachrymatory factor synthase can be turned off in the onion, which produces onions that don't make us cry, but they are not readily available.

Shallot

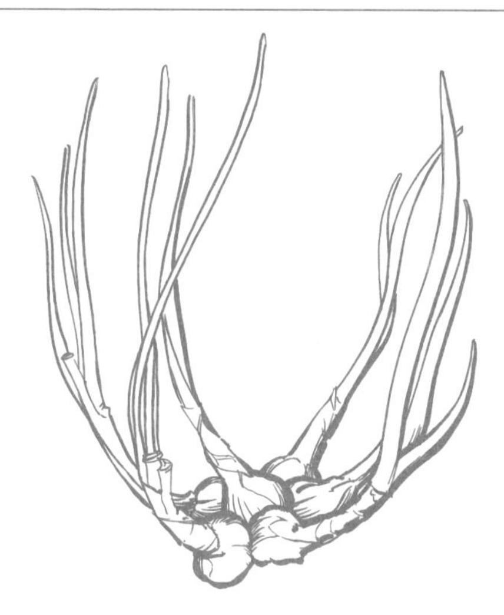

***Allium cepa* Aggregatum Group**

Shallot, Potato onion, Multiplier onion

Multiplier onions are botanically almost identical to their parent, *Allium cepa*, but when a bulb is planted, they will form a small bunch of several onions in the ground. This comes from vegetative propagation within the parent bulb. On the basal plate, new buds form at the axis of the storage leaves. These buds become new bulbs, seemingly splitting out from the mother bulb and forming a ring around her. The new bulbs grow with their tunica and will eventually put down their own roots, although they are usually harvested before this occurs.

Allium sativum

Amaryllidaceae

Garlic

Garlic is another edible Allium, originating in the eastern Mediterranean and cultivated mainly for its edible bulb, although all parts can be eaten. Like its close relative, the onion, it is a tunicate bulb, but unlike the onion, the bulb develops underground with only the leaf blades visible. It is a biennial, grown as an annual, but most modern garlic cultivars are sterile, reproducing only through vegetative means, splitting into many cloves from one parent clove.

The basic anatomy of garlic is the same as the onion, with a basal plate or flattened and modified stem producing roots below and leaves above. The lower half of the leaves are thickened for storage and the top half are hollow blades that photosynthesise. The leaf blades of garlic are more flattened than those of onions. Garlic bulbs are composite – within the tunica of the parent plant are several smaller tunicate bulbs, referred to as cloves.

Garlic begins life as a clove, which it produces from the axillary bud of the storage leaves of the parent plant, similarly to multiplier onions, *Allium cepa* Aggregatum Group. A clove consists of a small central bud surrounded by a single thickened storage leaf and covered in

The botany of seed sowing

A great proportion of the kitchen garden is propagated annually from seed. A seed contains the embryo, which will give rise to the new plant. There are two main types of seed: endospermic and non-endospermic. Endospermic seeds such as corn, *Zea mays*, store food for the embryo in a specialised tissue called the endosperm. Non-endospermic seeds such as French beans, *Phaseolus vulgaris*, store reserves in the cotyledons or seed leaves. In both cases, the food reserves and embryo are tucked safely into the seed, which is covered by the seed coat or testa.

The seeds generally leave the plant while in a state of dormancy, meaning they will not grow as soon as they hit the soil. Instead, the components inside the seed are held in stasis, and respire very slowly to stay alive. This respiration uses some of the food reserves, and this is why seeds do not store forever. This induced dormancy has many benefits, including that the seeds do not germinate until the conditions are correct. Many seeds are produced at the end of the growing year, and if they germinated straight away would be growing in a cold, dark winter, so it is beneficial to wait until the warmth and light of spring. Dormancy also allows seeds to be transported, often by animals, until they reach a new growing spot. This evolutionary tactic allows plants to spread and dominate, often finding better places to thrive. In perennial plants, it also means the seedlings are not in direct competition with the parent plant for light, food and water.

The seeds generally leave the plant while in a state of dormancy, meaning they will not grow as soon as they hit the soil.

Most seeds need water, oxygen and the right temperature to start germination, although some also require complete darkness, and others need light. Seeds have many dormancy tactics that need to be overcome before they can germinate. Some of the most common include the testa being impermeable to water or restricting

gaseous exchange, as in beetroot, *Beta vulgaris*. Some have germination inhibitors in the fruits around the seed, such as in the flesh of the tomato, *Solanum lycopersicum*, which must be fully removed for the seeds to germinate.

Often, all that is required to break dormancy is for the seed coat to be thinner or removed, a process that happens naturally, either within the digestive tracts of animals, through the action of soil micro-organisms, or through the weathering of the cold winter with freezing and thawing. In the kitchen garden, the seeds can be scarified to mimic the natural process, by simply scratching on sandpaper, or nicking through the top layer to allow water in. In some cases, just soaking in warm water will remove the waxy outer layer and allow gas exchange.

Once all the right conditions have been met, with water and oxygen entering the seed, and the grower providing the right temperature for the particular crop, the seedling starts to develop. Initially, the radicle or primary root, which has been tucked inside the seed, makes its way down into the earth and sources water and nutrients. Then the primary shoot or plumule reaches upwards.

The plumule consists of three parts. First, the cotyledons or seed leaves, which can be 'epigeal' and appear above the ground, or 'hypogeal' and never break the surface. The first leaves that appear above the ground are the true leaves. When pricking out it is important to not damage these seed leaves. The part of the plumule below the cotyledons is the hypocot, and the bit above them is the epicot.

Often, all that is required to break dormancy is for the seed coat to be thinner or removed.

In monocot plants, such as *Zea mays*, the radicle is protected by the coleorhiza. This grows down first, followed by the radicle, ensuring it is not damaged. The cotyledons also have a protective layer, known as the coleoptile.

After this point, the plant has true leaves to start photosynthesising and creating its own energy, plus a root to anchor it and search out water and nutrients.

Garlic bulb

Garlic plant

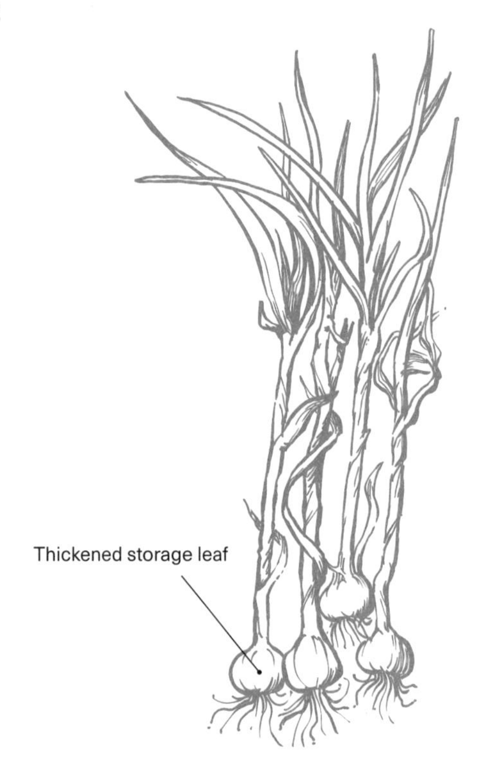

a protective sheath, or tunica. The cloves are produced within the skin of the parent and yield a composite bulb. After harvesting, the cloves are either skinned and eaten or planted to become parent cloves.

There are two distinct varieties of garlic that are planted in the kitchen garden, namely *A. sativum* var. *ophioscorodon*, also known as hardneck garlic, and *A. sativum* var. *sativum*, which is referred to as softneck garlic.

Hardneck garlic produces a flower spike, known as a scape, but it does not form seeds. Instead this false flower spike produces small bulbils, which can be planted and will grow into a garlic bulb after several years. To produce a scape, the parent clove must go through vernalisation, or a period of cold, which will trigger the end of dormancy. Hardneck garlic needs around two months of temperatures below 10°C (50°F). Plants generally need vernalisation if their wild ancestors experienced cold winters, and and therefore had to stay dormant during this period. Once a certain length of cold has passed, hormones are released to tell the plant that spring is here, the conditions are right and growth or the next stage of development should begin. Hardneck garlic will not produce cloves if it does not produce a scape, instead forming one large clove, as the signal to reproduce via axillary buds is not received. The scapes themselves are delicious and should be harvested once they begin to curl at the top, to stop the garlic putting too much energy into producing a flower rather than into bigger cloves.

Softneck garlic does not produce a flower spike, and therefore does not need vernalisation. In comparison to hardneck, it produces more but smaller cloves.

Ananas comosus

Bromeliaceae

Pineapple

The pineapple was first domesticated in the tropical climates of South America, and as such, still only thrives in hot countries, or under glass where outdoor temperatures are not so high. *Ananas comosus* is a monocot and a member of the Bromeliaceae. It displays many typical traits of bromeliads, such as collecting pockets of water at the base of the leaves where they grow closely together, known as phytotelmata. Pineapples also have weak roots as do other bromeliads, because they are descended from epiphytic plants, those with aerial roots.

Pineapples are one of many crops that produce their young via parthenocarpy, where no fertilisation occurs, and the resulting fruits

Pineapple

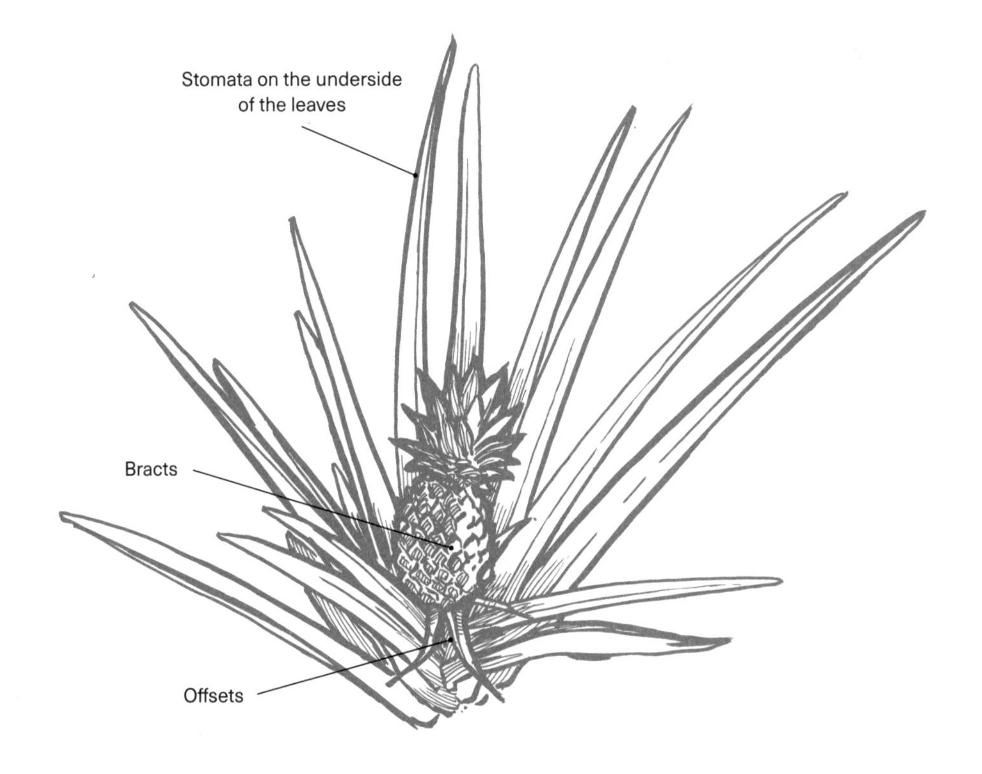

are sterile. In the wild, fertilisation does occur, with hummingbirds as the main pollinator. The fertilised fruits contain lots of very hard seeds – not ideal in a fruit crop for humans who want a quick snack – so the parthenocarpic fruit is advantageous to the grower. Cultivated pineapples are propagated via vegetative propagation, with the leaves at the top of the fruit and the offsets below all able to produce roots and be used as slips (cuttings).

The fruit of the pineapple develops from around 200 individual flowers, each with a bract, which will eventually become the sharp point on the external rind. Each of these flowers becomes a berry fruit, deriving from the ovary, but then the receptacle also fuses into the final pinapple, meaning this can be classified as an accessory fruit. Thankfully, to give it one final classification, a compound fruit with a central stalk is known as a sorosis.

The calyx of each flower folds over as the fruit develops and creates one of the bumps on the rind, and the bract dries to become sharp and point outwards. If the rind is peeled off, lots of dark indents can be observed, which are the remains of the top of the ovary and rim of the receptacle. The central core is the softened stem, which is full of indigestible fibrous tissue, which was once the vascular system. As with most monocots, this tissue is spread throughout the core, rather than just around the outsides as would be found in dicots.

Eating pineapple can often leave a tingling sensation on the tongue, from the enzyme bromelain, which is found in the fruit and stem. The purpose of this enzyme is to digest protein, and is the reason pineapple is often used as a tenderiser in cooking. Within the pineapple, this enzyme aids the ripening process.

Having to grow where there is little water available for the roots has led the pineapple to have several interesting adaptations. The leaves show some typical water-retaining features, with a layer of colourless cells whose only purpose is to swell up and retain water. The leaf has very few stomata and the majority of these are on the underside of the leaf. When the stomata are open for gas exchange, they are also a site of water loss via evapotranspiration. To reduce this even further, the stomata are surrounded by mushroom-shaped cells originating from the epidermis, called trichomes. These trichomes trap moisture, creating a humid environment around the leaf, reducing transpiration as the water concentration gradient of the leaf is equal both internally and externally. The trichomes also give the underneath of the leaf a silver sheen, which increases reflection of light and reduces leaf temperature further, again lowering water loss.

Another adaptation is the way the pineapple photosynthesises. Most plants open their stomata during the day to take in carbon dioxide for photosynthesis, but in hot conditions, this rapidly leads to water loss. Pineapples use CAM photosynthesis to avoid daytime water loss. CAM is short for Crassulacean Acid Metabolism and involves opening the stomata at night to take in carbon dioxide. The gas is then stored as malic acid in the cell vacuoles until daytime, when it is carried back to the chloroplasts, converted back to carbon dioxide, and photosynthesis takes place.

Apium graveolens

Apiaceae

Celery, Celeriac

Apium graveolens grows wild along moist coastal places and originated in the Mediterranean. It is a biennial, which in its second year produces a large flower stalk culminating in axial umbels. Its family, the Apiaceae, was once called the Umbelliferae, and all plants within it produce similar types of flower that originate from one place and form a flat head made of lots of small flowers on short pedicels or stalks. Wild celery has hollow stems, which are poisonous and taste very bitter, a trait that has been bred out in the cultivated varieties. All of the plants related to *A. graveolens* have a strong and particular scent and taste – in fact the specific epithet *graveolens* means strong-smelling. The molecule that produces this smell is called sedanolide and is a volatile molecule, which doesn't break down on cooking, which is why the taste of celery remains strong in soups and stews. All of the *Apium* plants have a large fleshy tap root, another common trait of plants in the Apiaceae family.

Apium graveolens var. *dulce*

Celery

This variety of *Apium graveolens* is the most commonly grown type. Often when served, the part of celery we eat is referred to as a stem, but botanically it is a petiole or leaf stalk. The petiole starts at the base of the plant, originating from celery's short and round stem, and then at the top turns into a smaller branching section of petiole called the rachis, culminating in a pinnate

Celery

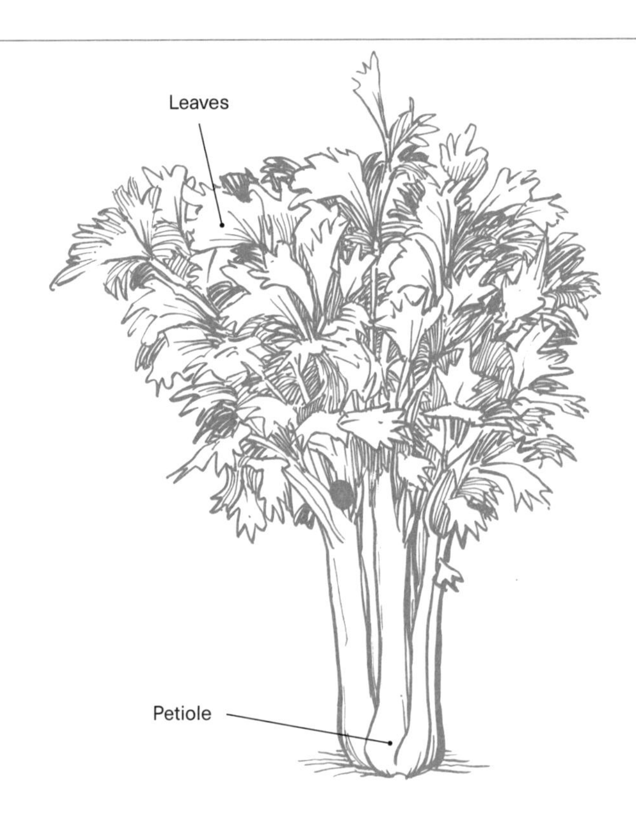

compound leaf. If the petiole is observed closely, there are many telltale signs that it is a leaf stalk, not a stem. The petiole is a crescent shape, which means it is symmetrical bilaterally, unlike a stem, which is usually symmetrical radially. Another sign is that the petioles are arranged in a Fibonacci sequence around a central axis, which is a common arrangement of leaves.

Celery has a fleshy, swollen petiole that comes in a variety of colours, from the typical green to yellow and red. The taste of celery can be slightly bitter owing to the presence of apiin, but the concentration of this can be reduced by blanching the petioles in a trench that is backfilled while the plant grows, stopping sunlight hitting it. This is why celery is often referred to as trenching celery, but there are modern cultivars that are self-blanching as they grow tightly and stop the light reaching the centre.

Another aspect of celery is its stringiness. There are two types of string that cause this. One is in the inside of the petiole and is the vascular bundles of phloem and xylem. But the more crunchy strings on the outside are formed of stacks of collenchyma tissue. Unlike other strong, structural tissues that are dead, collenchyma is a living tissue formed of cellulose and pectin. In the wild, these structural columns stop the petioles bending and breaking in the coastal winds, as does the C shape of the petiole, which has a similar effect of strengthening the plant.

Celeriac

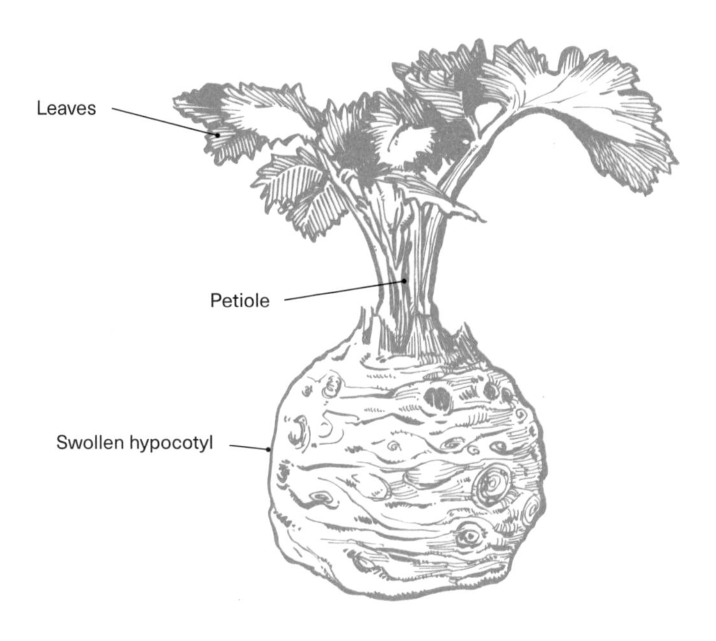

Apium graveolens var. *rapaceum*
Celeriac, Turnip-rooted celery
This variety of celery is grown for its large, swollen hypocotyl, the area of stem under the cotyledons, the seed leaves, where the plant stores energy in the form of sucrose during its first year to help produce a flower in the second year. Celeriac is commonly a winter vegetable as it is fairly hardy, with white flesh and brown, wrinkled skin, growing half above the ground and half under it. Celeriac has bitter-tasting petioles and leaves, which are a darker green than celery and usually discarded. There is one variety of celery, *A. graveolens* var. *secalimum*, that is cultivated for its leaves, and is known as leaf celery or Chinese celery.

Arachis hypogaea

Fabaceae
Peanut, Groundnut, Goober, Monkeynut

Going by many names, *Arachis hypogaea* is commonly looked upon as a nut, but botanically speaking a nut is a dry, hard fruit whose case does not split open to disperse the seed. Instead, the peanut fruit is classified as a legume. Originating in South America, this warm-season annual is grown for its seeds and oil, and only really thrives in hotter countries where it gets a full six months of higher temperatures.

The peanut plant is around 50 cm (20 in) in height, but there are two forms, erect and prostrate, although most commonly it

The pigeons have attacked some of the crops, but we are leaving them in to grow back.
Sweetcorn 'Glass Gem'

Peanut plant

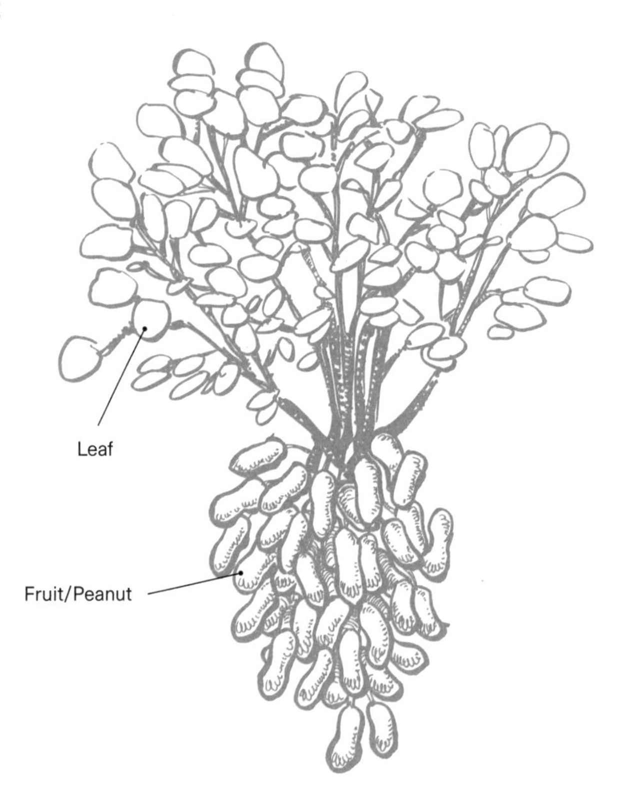

is the erect form that is grown. It has four pinnate, hairy leaves, which grow as two sets of opposite leaflets. The yellow flowers are sessile, meaning they have no pedicel or flower stalk, but instead have a long, fused calyx tube, which is formed from five fused sepals. The flowers only last for a day, but thankfully are self-fertile, which means the new generation are clones of the parent, apart from very rarely when an interested bee visits the flower and spreads a few pollen grains. Once the flower has been fertilised, the petals drop, and this is where the peanut starts to show some interesting behaviour.

Arachis hypogaea is geocarpic, which is a rare form of reproduction where the fruit forms underground, with *hypogaea* literally meaning under the ground. This helps defend the precious new growth from the elements above ground. Once fertilisation has occurred, the base of the ovary elongates and develops into a long structure called a peg or stipe, which originates from an area of meristematic cells just below the receptacle of the flower. Fertilisation triggers the production of amyloplasts in the apical region of the peg. Amyloplasts are starch-filled cell organelles, that tell the plant which way down is, as they are heavier than the surrounding cytoplasm, and sink to the bottom. Starchy amyloplasts are called statoliths and help sense balance.

This is one method the plant uses to become geotropic, responding to gravity, and enables the process of taking the fertilised ovary towards the ground. Once the ovary is safely buried 2–5 cm down (0.8–2 in) the geotropic properties cease and it begins to grow horizontally. Although structurally a stem, once underground the peg starts to act like a root, turning its smooth epidermis, laden with stomata and lenticels, into an area with small hairs that take up moisture and nutrients from the soil. When underground, the ovary finally starts to develop into a fruit.

The fruit itself has a tough, fibrous shell that develops from the ovary, and is useful for protecting the underground seed. Within it, the seed has a papery, edible coat that encloses two enlarged cotyledons, which are full of stored proteins. As it is the cotyledons that are the food storage for the seed, not the endosperm, these seeds are described as non-endospermic. If the two halves of the seed are split open, a small, white tuft is observed, which will develop into the roots and shoots of the new plant once conditions are correct.

Armoracia rusticana

Brassicaceae

Horseradish

This perennial herbaceous plant is a pungent root crop that is rarely eaten by the mouthful, instead being grated and added with caution to meals. Appearing in the Brassicaceae, *Armoracia rusticana* or horseradish has the mustard oil pest defence exhibited by many members of this family. The strong flavour is not present until the root is grated or damaged, then its unmistakable taste is released, theoretically to stop herbivores devouring it. Unfortunately for the plant, humans take pleasure in the heat.

Ironically, it is this love of the heat of horseradish by humans that has led to its spread around the world from its origins in Russia and Eastern Europe. It is known as a synanthropic plant, which means it is a wild plant that benefits from its association with people. In this case, horseradish is only found in places cultivated by humans, usually where they have disturbed its large tap root and left several pieces in the ground, where its root cuttings produce new plants.

Horseradish is mainly propagated by root cutting, meaning most plants are clones of each other. Once thought to be sterile, this is not the case – it can produce viable seed but has low fertility. All the pollen produced is fertile, but the plant cannot easily recognise it and rejects its own pollen. When fertilised with its own pollen, the seed produced is not viable. To produce new cultivars of horseradish, breeders have to find two plants that are not identical clones and cross-pollinate these, which gives a good percentage of viable seed.

The plant itself has a large, swollen tap root that is full of carbohydrates. Being full of solids means the root of horseradish is very frost-tolerant and can be harvested year round. Above ground, it has coarse, long leaves, often with roughly toothed margins, although some smooth-margined types exist. Some types also exhibit variegation, which is a genetic mutation within the leaf in which two different genetic kinds of the same tissue exist at once, one producing green areas and one an off-white colour. This makes the plant a chimera.To continue variegation, the plant must be propagated vegetatively, through root cuttings in this case, as the seed produced by horseradish is heterozygous – taking genes from both parents and not guaranteeing a variegated offspring.

Asparagus officinalis

Asparagaceae

Asparagus

Asparagus spears are one of the signs that spring is here. Harvested when 15–20 cm (6–8 in) high, they are the shoots of a monocot, herbaceous perennial that will eventually reach nearly 2 m (6 ft). Cultivation of asparagus most likely began in the eastern Mediterranean around 2,000 years ago and was a popular harvest with the Romans. It is a maritime plant and adapted as a halophyte, meaning it can tolerate salty and harsher conditions than many other plants, but for this reason it also dislikes extremely wet soils.

The shoots appear from the underground organ, called a rhizome, which is in fact a modified stem that stores nutrients and food over the winter to allow the shoots to appear every spring, with larger, older rhizomes being able to produce larger spears. The spears grow extremely fast, up to 3 cm (1 in) a day when conditions are favourable, at 11–25°C (51–77°F). At the top of the spears are tightly packed, small developing branches, which are protected by small triangular scale leaves that stop any damage as the spears break through the surface. The spears can be purple and green, but it is also possible to produce white asparagus. All asparagus spears are white until they reach the surface where they begin producing chlorophyll and photosynthesising, so heaping soil onto the crown means that the spear has further to grow until it produces chlorophyll.

The older the spear is when it is harvested, the woodier and tougher it is. As with all monocots, the vascular bundles are spread throughout the stem, and the vessel walls thicken with lignin as they age, as do the sclerenchyma cells just under the skin of the spear that give support to the

Asparagus

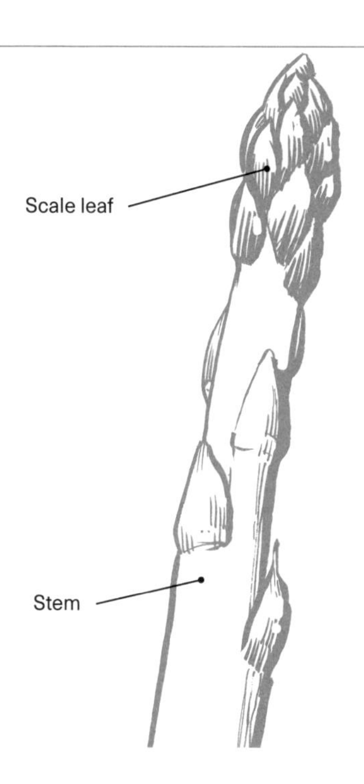

extremely tall stem. Once harvested the enzymes responsible for this process continue working, meaning asparagus can actually get tougher if stored for long.

Shoots that are not harvested grow upwards and create finely dissected foliage that resembles that of a fern. This foliage is not made up of true leaves, but instead cladodes or modified stems. This adaptation helps to reduce transpiration in the harsh maritime conditions where asparagus grows wild. The cladodes are the main organ involved in photosynthesis and still contain stomata, but many fewer than a leaf. The true leaves of the asparagus are in fact the tiny, triangular scale leaves found protecting the tip of the spear, and up and down the stem.

Once fully grown, asparagus will produce flowers and as it is dioecious, there are separate male and female plants. The female plants will produce small red berry fruits upon fertilisation. There is a modern tendency to breed all-male cultivars, as they tend to give fatter spears because they do not have to expend energy producing fruits and can therefore store more in the rhizome over winter.

Beta vulgaris

Amaranthaceae

Beetroot, Chard

Beta vulgaris has an unclear history, but is likely to have originated in coastal regions in Eurasia – certainly many of the plant's adaptations have given it the ability to thrive in the harsh salty conditions of coastal regions. *Beta vulgaris* gives us several different important edible subspecies, including beetroot, sugar beet and chard, all

The botany of vegetative propagation

Although most of the new plants grown in the kitchen garden are from seed, and therefore the product of sexual reproduction, there are some that are created from vegetative propagation. This form of multiplying plants involves using existing material from a mother plant to grow offspring that are clones – genetically identical. The production of clones does occur in the wild, with many plants spreading via horizontal stems or 'stolons', such as the strawberry, putting down adventitious roots at nodes and creating new plants. In evolutionary terms, this helps the plant to dominate the growing space, out-competing other species for nutrition and water. Clones do have some drawbacks, as they may contain diseases from the parent plant and will share other weaknesses.

Vegetative propagation revolves around parenchyma cells, which are undifferentiated and can form new adventitious roots. New roots can also form from callus, a form of parenchyma cell, which is produced when a plant is wounded. These cells can differentiate to become roots, as well as shoots, and can therefore produce a whole new plant, which, given time and the right conditions, can survive independently from the parent. This process is also reliant on the hormone auxin, which promotes the growth of new roots. It is commonly produced in the apical buds and leaves and is transported down through the plant.

In the kitchen garden, vegetative propagation is most common in producing new fruit bushes and trees, either via grafting or taking hardwood cuttings. This guarantees that the offspring will produce the same tasty fruit that the parent does, because most fruit is cross-pollinated, whereas the offspring from seed can show huge differences from the parent plants. These cuttings are generally taken as hardwood cuttings, or woody stems from the parent. They contain apical buds, so have an auxin store, which as it flows down will gather at the cut surface at the bottom of the hardwood cutting and promote its root growth. The energy for root establishment comes from stored carbohydrates, as hardwood

In the kitchen garden, vegetative propagation is most common in producing new fruit bushes and trees, either via grafting or taking hardwood cuttings.

cuttings are generally taken once the leaves have fallen. Adding heat, water and oxygen improves the chances of successful root formation, and production of a new plant.

Auxin is produced in greater concentration by juvenile parts of the plant, and it is this trait kitchen gardeners are taking advantage of when they multiply plants with softwood cuttings. Crops such as tomatoes, peppers and sweet potatoes can all be multiplied by removing soft new shoots from the parent plant and putting them into either water or free-draining compost in a warm environment, where they will form new roots in a matter of days. As well as a good supply of auxin, they also have leaves that continue to photosynthesise and produce energy.

Some vegetative propagation comes from simply dividing an existing storage organ, such as in rhubarb.

Some vegetative propagation comes from simply dividing an existing storage organ, such as in rhubarb. Storage organs, like rhizomes and tubers, generally get larger year on year, and have several buds or growing points. A section of the storage organ can be split away from the parent, as long as it has this growing point. Ideally, it also comes away with roots, and can then grow away easily. In storage organs such as potato tubers, the potato can be cut into sections, as long as each one has an 'eye'. These sections will then produce roots, and grow into new plants given the correct conditions.

likely to have evolved from sea beet, *B. vulgaris* subsp. *maritima*.

Beta vulgaris subsp. *vulgaris*

Beetroot

Most vegetable plots will be home to a row of beetroot, typically a red, globular crop cultivated for its edible storage organ. Although classified as a root crop, the swollen section of the beetroot is actually a combination of root and the hypocotyl, which is the section of the emerging seedling just below the cotyledon and above the young roots. Most globular beetroots in the kitchen garden are predominantly hypocotyl and rich in carbohydrates.

Cut open a beetroot and you will be greeted with obvious rings, particularly noticeable in cultivars such as 'Chioggia' where they are alternately red and white. These are growth rings of vascular bundles and storage cells. The thinner, lighter rings are conductive tissues made of xylem and phloem to move water and organic compounds around the plant, while the broader, darker rings are storage tissues where sugars are stored as carbohydrates. Each ring has an active cambium layer producing more cells, whereas most flowering plants only have one cambium layer just under their outer surface.

The taste of beetroots is often described as sweet and earthy, and this storage organ consists of 6–10% sucrose. Farmed sugar beet *B. vulgaris* subsp. *vulgaris* Altissima Group contains 18% and produces a large amount of the world's sugar. The earthy flavour comes from a compound called geosmin, which is thought the beetroot produces to help with protection against salty seaside conditions. Soil organisms such as bacteria and fungi also produce geosmin, and this gives the 'petrichor' scent after it has rained on dry, warm soil.

Beetroot is a biennial crop, producing wind-pollinated flowers in its second year. When stressed it will bolt in the first year, but these seeds are never saved as this early flowering trait is not useful in the kitchen garden as the roots produced tend to be small and woody, not something to pass on. Over winter, the plants need a period of vernalisation of below 10°C (50°F) for 30–60 days to produce seed in the second year.

Beetroot

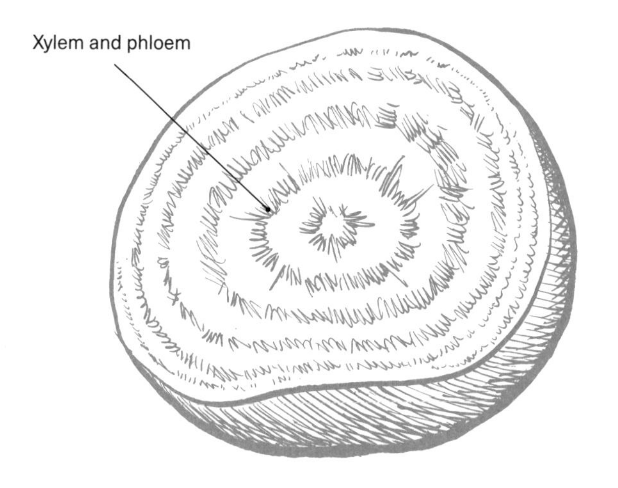

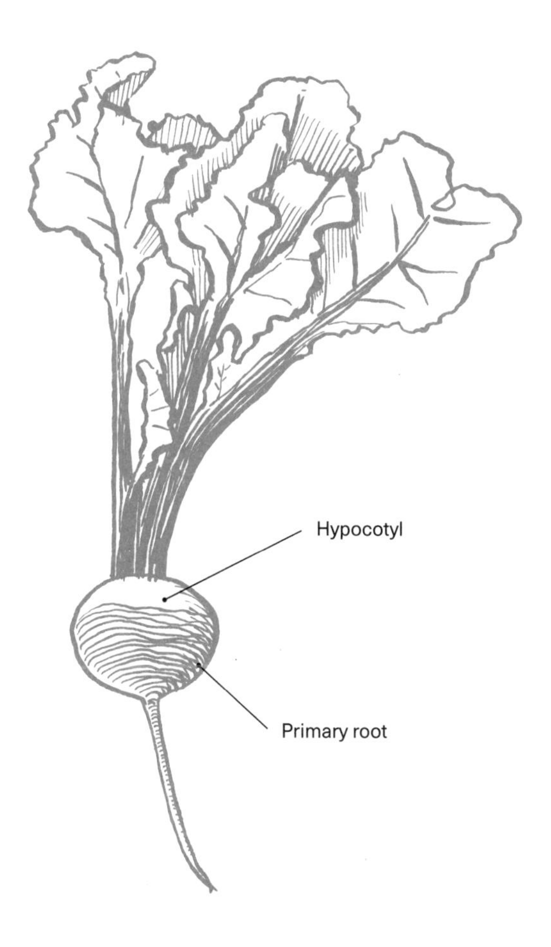

Chard

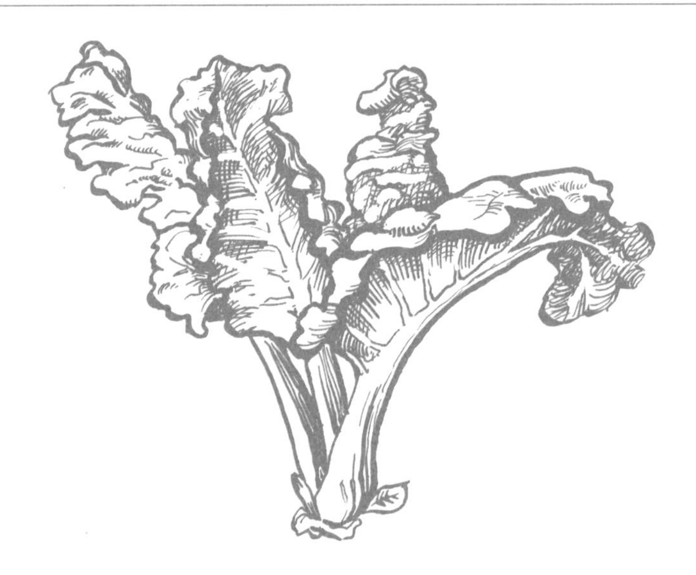

Swede

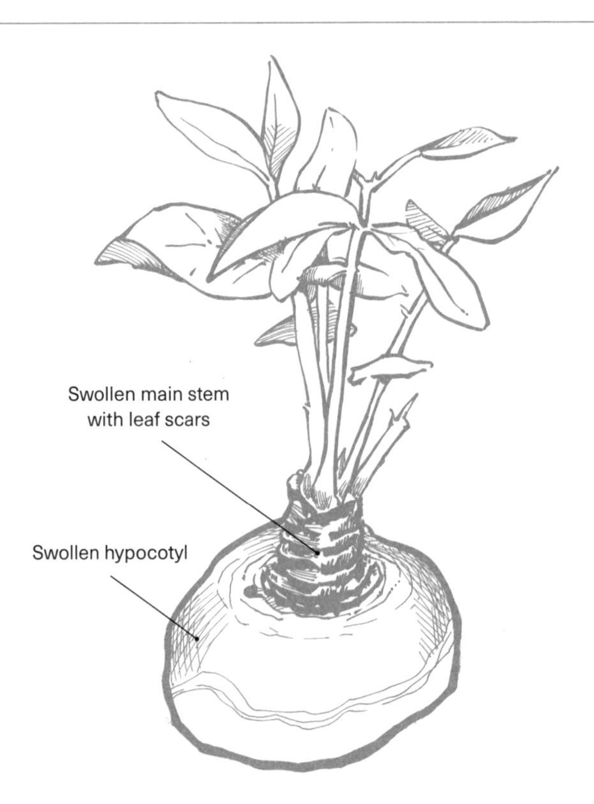

The flowers coalesce to produce fruits, which contain clusters of seeds and a hardened receptacle and perianth, known as the calyx. This coating provides protection and food reserves that are composed of the perisperm (remains of the nucellus, which enclosed the embryo) and a germination inhibitor. These seed clusters, known as glomerule or multigerm, lead to small groups of seedlings being produced. These need to be thinned for large beets, although they can be left for clusters of smaller beets. Breeders have developed monogerm varieties, which don't need thinning, particularly beneficial in large-scale beetroot cultivation.

Beta vulgaris subsp. *cicla*

Chard, Swiss chard, Silverbeet

The leaves and stems of beetroot are edible, colourful and delicious, making them a fantastic secondary harvest. But the cousin of beetroot, chard, is grown purely for its edible petiole and leaves. Unlike beetroot, it does not have a swollen root, but instead produces a swollen mid-rib and large, abundant leaves. The leaves are often puckered and have a slight waxy texture. This wax is an adaptation from its time growing on the shoreline, where the coating would reduce the transpiration of water, and stop damage from the salty atmosphere.

The leaves can be harvested regularly and the plant will continue growing through milder winters, but is still a biennial plant so will go to seed in the second year.

Brassica napus

Brassicaceae

Oilseed rape

This species is closely related to *Brassica rapa* – in fact it is thought that *B. napus* originated as a hybrid between *B. rapa* and *B. oleracea*. *B. napus* is tall plant, growing to 1m (39in) with glaucous leaves and is cultivated for its yellow flowers, which produce oil-rich seeds that are processed to give a vegetable oil.

Brassica napus Napobrassica Group

Swede, Rutabaga, Neeps

A swollen root crop, the swede often finds itself compared to the turnip, which from a distance is an understandable mistake, as these two swollen root crops in the brassica family have a similar taste. The swollen part of the swede is the hypocotyl and main root, as with the turnip, but in the swede part of the main root also swells and contributes to the enlarged root. This area can be identified as there are several horizontal ridges on the top of the swede, which would have been leaf scars. The leaves of the swede are hairless, and the flesh is generally yellow, although occasionally white. It is also a much more hardy crop than the turnip.

Brassica oleracea

Brassicaceae

Wild cabbage

Naturally occurring *Brassica oleracea* is known as the wild cabbage, which is native to coastlines all the way from the Mediterranean to Britain. It is found growing in harsh maritime conditions with its strong tap root burrowing deep into stony slopes. It is a perennial plant with waxy blue-green leaves, which are often seen in its cultivated relatives in abundance. Although not a showy plant, this wild brassica is the beginning point for so many brassica crops grown today, also known as cole crops.

From this simply formed perennial, over time, different elements have been bred or have mutated to enlarge and form a wide diversity of important edibles. All have the same elements, namely glaucous, edible leaves, a cruciform

flower with four petals and strong roots. Another aspect they all enjoy is a soil pH of around 6.0–6.5, which is slightly acidic, but it is commonly believed that they like an alkaline soil.

The benefit of alkaline soil is to counter one of the many garden enemies of the brassica – clubroot or *Plasmodiophora brassicae*. A soil-borne phytomyxea, a type of parasite, this affects the roots of brassicas, causing deformity and often leading to poor head development. Clubroot thrives in a soil pH of 5.7, which is very acidic, and is still active in the range 5.7–6.2, eventually being thwarted with a soil pH of 7.8. Brassicas will grow in soils with an alkali pH and it is a good way to rid the garden of this issue, but if clubroot is not present, the brassicas will thrive in slightly acidic surroundings.

These cultivated forms of *Brassica oleracea* are often cited as varieties, but here are classed as cultivar groups where applicable.

Brassica oleracea Alcephala Group

Kale, Borecole, Collards

Acephala Group is thought to be the closest relation to the wild cabbage and is commonly termed as kale. Cultivated for its leaves, it is a non-heading brassica whose lateral and terminal buds elongate during its first year, creating an often tall plant. Its leaves come in many forms, from dark green to purple, all covered

with a waxy cuticle, which stops water loss via evaporation, an important feature when growing in a maritime site. This slippery coating also prevents disease and pollution entering the leaves and in some cases can even make them too slippery for insects to lay eggs on.

Most kales are biennial, but some, such as *B. oleracea* var. *ramosa,* are short-lived perennials.

Brassica oleracea **Alboglabra Group**

Chinese kale, Chinese broccoli, gai laan, kai lan

Again cultivated for the leaves, the Alboglabra Group tend to have more tender leaves than the Acephala Group, and are used in many traditional Asian recipes.

Brassica oleracea **Botrytis Group**

Broccoli, Cauliflower, Broccoflower, Calabrese

These plants are cultivated for their edible, terminal inflorescence, which can be coloured anywhere from white for cauliflower to green for broccoli. The parts consumed by humans are the unopened flower buds, which are tightly packed together in a compound corymb (flat-topped) formation. They have short apices and short internodes, which helps to create the illusion of one large head whereas upon closer inspection, separate inflorescences can be seen. The mutation from the original wild plant occurs at the inflorescence meristem, which rapidly divides, giving much larger heads than in the wild plant.

Cauliflower

Cabbage

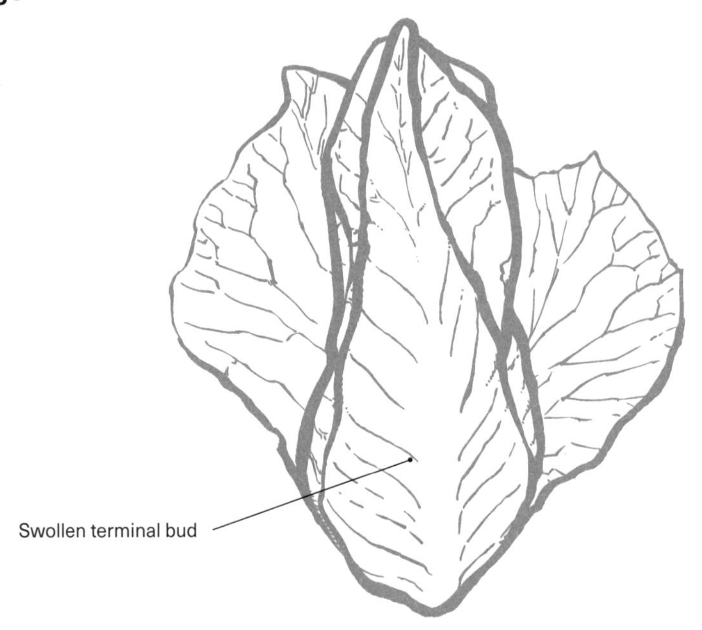

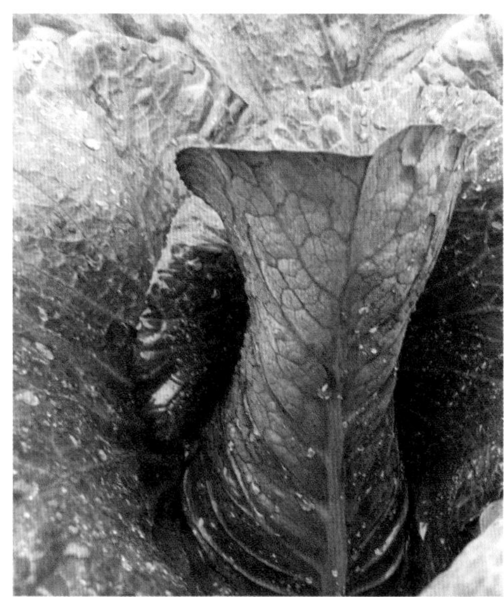

For cauliflower, the curds or head are surrounded by the leaves, which gives it its white appearance as they block the sun and no chlorophyll forms here; if the green leaves do not form properly and the curds are in direct contact with the sun, they will discolour and taste bitter. Cauliflower was bred in two distinct locations, which give us early and late types. The late types form curds when the temperature is 10–16°C (50–60°F) and are closely related to the wild cabbage, but some were bred in the more tropical conditions of India, and form their curds at 20–25°C (68–77°F), giving us a cauliflower that crops earlier in the season or in the middle of summer in some countries.

Calabrese heads are the blue-green colour, which is common in brassicas, and again are cultivated for their immature stalks and young flower buds, which form a dense head and often have fasciated stems, which are flattened and joined, but are still visible as individual stems, unlike those of cauliflower.

Brassica oleracea Capitata Group

Cabbage

Brassicas in the Capitata Group have a swollen terminal bud, which is harvested in the vegetative stage. It is surrounded by a mass of large leaves, which are tightly packed to form a head. This head comes in lots of different shapes: ball, pointed, conical and drumhead to name a few. They also 'head up' at different times of the year, meaning there is a cabbage for every season.

The core of the cabbage is the stout stem, which is enclosed in the outer leaves. It is easy to see when cutting open a head of cabbage the relationship between that and kale, with the whorls of leaves around a central stem, but the adaptations are also obvious, with the swollen terminal buds and shortened internodes. The new leaves grow inside the old ones, which often makes the oldest ones only good for protection and too tough for consumption by humans.

There are several colours of cabbage, including the standard green; white, which is

Kohl Rabi

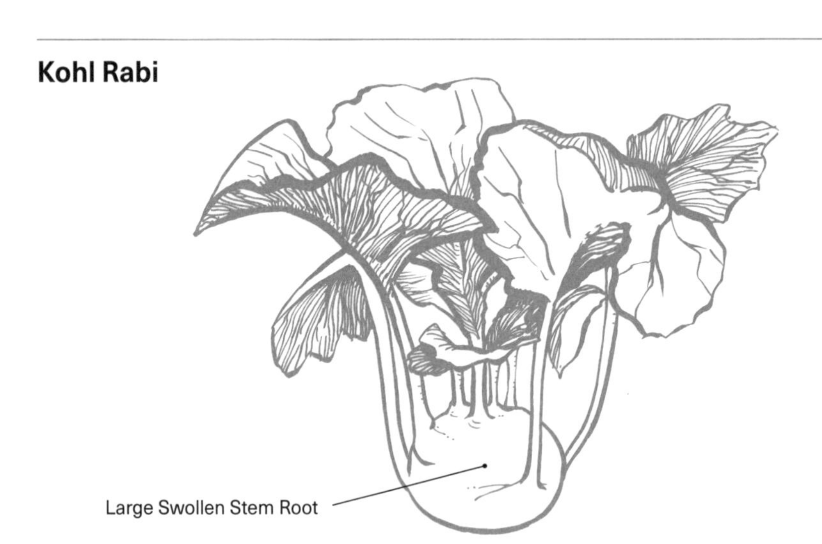

Brussels Sprouts

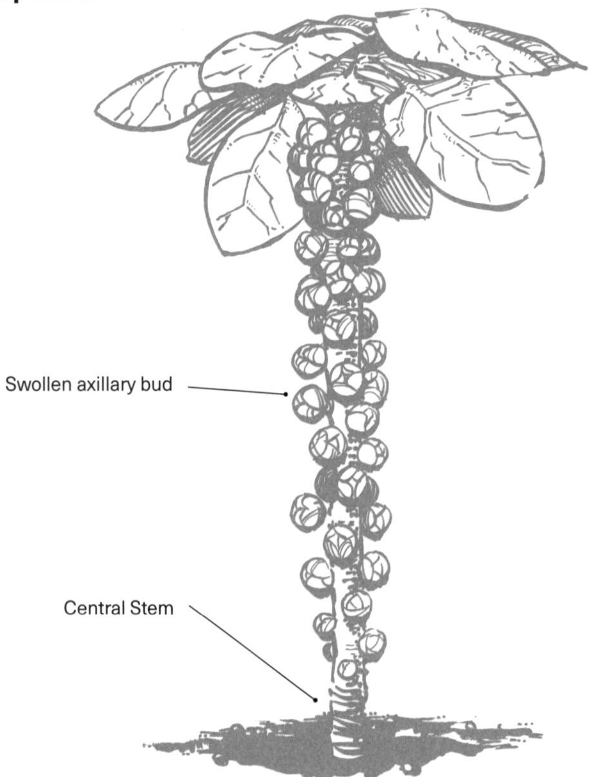

Sprouting Broccoli

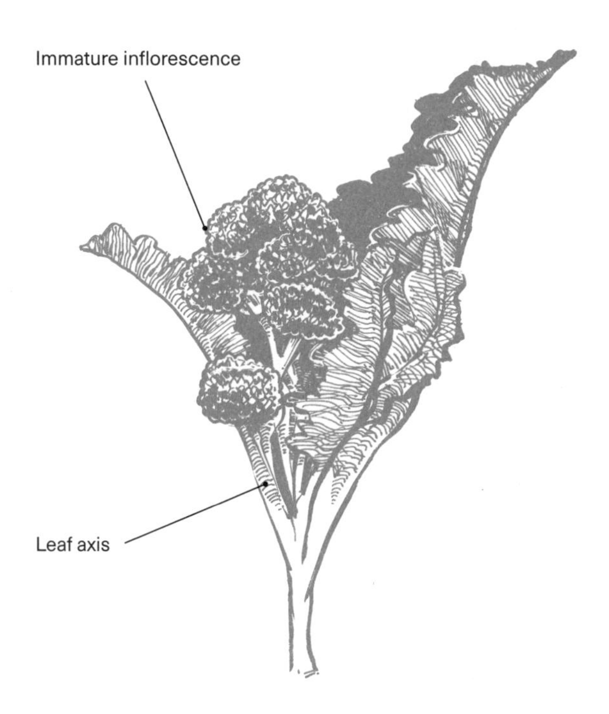

var. *alba*; and red, which is var. *rubra*. Savoy cabbages, var. *sabauda*, are the largest and hardiest cabbage, growing happily and slowly through the cold season. Savoys have the darkest green leaves, which closely resemble those of lacinato kale.

Brassica oleracea Gemmifera Group

Brussels sprouts

These relative youngsters to the family were bred around 500 years ago and have enlarged axillary and terminal buds. The structure of the sprout is similar to that of the cabbage, with the swollen pedicel in the centre instead of the large central stem. The sprout plant itself is often tall and has the internode proportions of most kales. As well as the axillary buds, the terminal bud is also swollen and surrounded by leaves, being known as the sprout top when harvested. The sprouts swirl around the main stem and mature from the bottom up, although some new F1 cultivars have been bred to mature at the same time for ease of picking.

Brassica oleracea Gongylodes Group

Kohl rabi

This mutation of the wild cabbage plant is generally found in root crops bed in the kitchen garden, but like many others, it is not a true root crop. Instead it has a large swollen stem, which sits predominantly above the soil. The lateral meristems, which are found on the sides of the stem for thickening to produce a support system for the plant, produce much more than in the wild form, causing this swollen area. This lateral meristem can also produce woody tissue for

Turnip

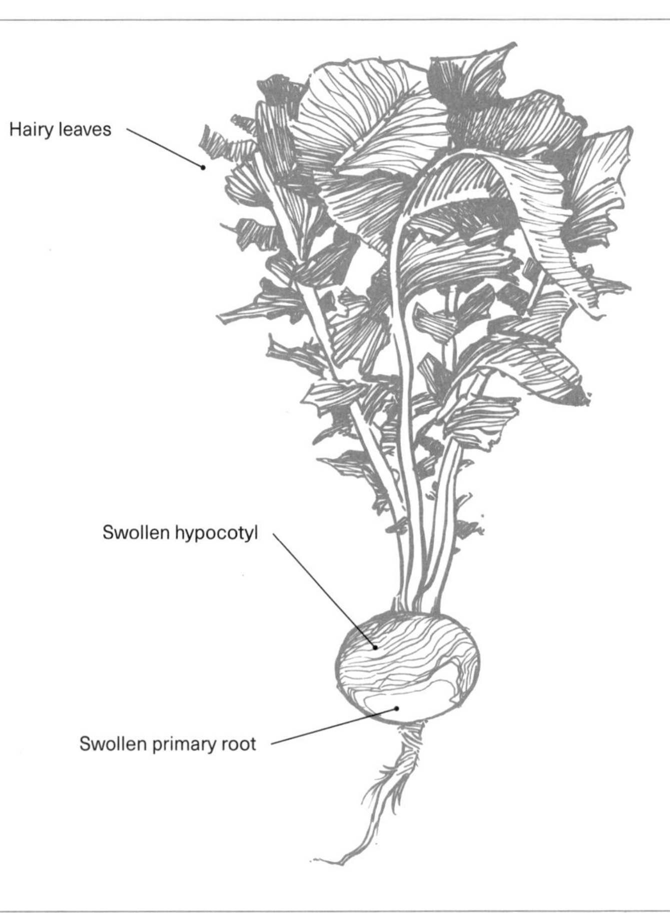

Pak choi

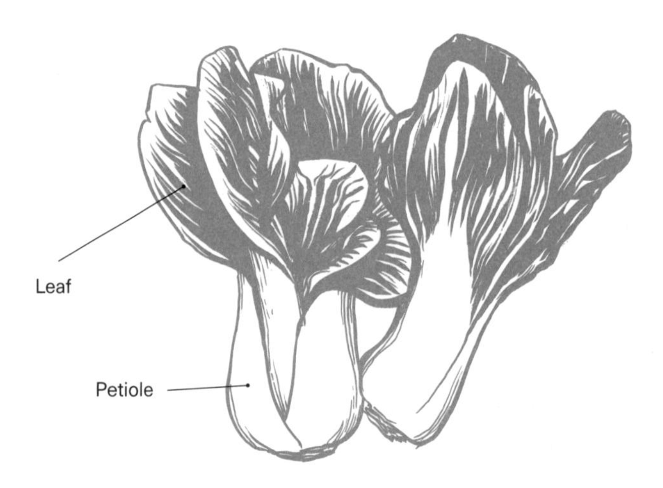

protection, and occasionally creates a thin layer on the outside of the kohl rabi, especially when older or damaged.

Brassica oleracea Italica Group

Purple sprouting broccoli, Sprouting broccoli

Harvested for its immature inflorescence, the young buds of sprouting broccoli form on a relatively long stalk. These inflorescences are much looser than those in the Botrytis Group and have lots of smaller heads. They are arranged as those in the Gemmifera Group, with the flowers growing down the stem in the leaf axis along with one larger terminal flower.

Brassica rapa

Brassicaceae

Members of the brassica family are often quite promiscuous and interbreed readily, which makes collecting seed that is true to type quite hard work, and they also show great variety within one species, as is the case with *Brassica rapa*. As with *Brassica oleracea*, *B. rapa* are primarily split into cultivar groups. The wild plant is small with a rosette of leaves and small yellow flowers with four petals.

Brassica rapa Rapifera Group

Turnip

This group of *B. rapa* has a fleshy root, which consists of the swollen hypocotyl and top part of the true root. This becomes swollen via lignified secondary xylem, which means the turnip can become a little woody. As a root crop, it most commonly has white flesh with purple and white skin, although it can also have yellow flesh. Some turnips have been cultivated for their green leaves, which are slightly hairy on the underneath and always edible, but more so in the turnip top varieties.

Brassica rapa Chinensis Group

Bok choi, Pak choi, Choysum

Cultivated for their edible leaves and petioles, these vegetable crops form a loose rosette of leaves with swollen midribs that are white or pale green. The leaf blades are generally dark green or purple and slightly shiny.

Brassica rapa Pekinensis Group

Chinese cabbage

The Pekinensis Group consists of heading vegetables, with large, densely packed leaves forming a barrel-shaped cabbage. The midribs are broad, although narrower than in the *B. rapa* Chinensis Group, and the leaf ribs are curled and pale green.

Capsicum spp.

Solanaceae

Peppers

Originating in the tropics of South America, *Capsicum* species are cultivated by humans for their fruits, generally referred to as peppers. Given their origins, they are not hardy plants, and will not live through a frost, preferring the heat and growing best with day temperatures of 27–32°C (80–90°F). They all produce complete cymose flowers. Although mainly self-pollinating, they will cross-pollinate, with bees, ants and thrips carrying pollen from plant to plant.

The fruit produced is a fleshy berry, which varies in colour depending on the species and cultivar, but all start off green with chlorophyll pigment. As they ripen, this compound disperses and is replaced by other compounds, such as carotene for orange peppers. The fruit itself often retains its peduncle when harvested, and the calyx forms a disc shape at the top of the fruit, which is cut out when eaten.

The fruit Is indehiscent, meaning it does not split to disperse its seeds, instead depending on animals to digest the fruit whole and spread the seeds. Oddly, the fruits contain the alkaloid capsaicin, which is a deterrent to most mammals, but birds can happily eat this compound with no ill side effects and spread the seed far and wide.

The capsaicin is produced in the placental tissues in the centre of the hollow fruits, where the seeds are attached. The capsaicin itself binds to receptors in the mucous membrane of the mouth, which are associated with heat and physical pain, which is why this alkaloid causes a burning sensation. If consumed often, the receptors are depleted, and a tolerance is built up. Bell peppers contain a recessive gene, which means they do not produce any capsaicin and are therefore sweet with none of the heat.

There are several species of *Capsicum* that are often cultivated by humans for consumption.

Capsicum annuum

One of the most commonly cultivated peppers, this is an annual plant, which bears its single, white or pale green flowers, with stunning blue anthers, in the leaf axils. The fruits produced grow either erectly or pendulous on the plant, are generally red or yellow when ripe, around 10cm (4 in) and have a thick flesh.

All bell or sweet peppers find themselves in the species *C. annuum*. The pungency or heat varies greatly within the cultivars, but it is still one of the milder species. Despite the wide variety of shapes and sizes of fruits found within this species, several attempts have been made to group them, and there are five cultivar groups that often appear:

- Grossum Group – bell peppers with large, mild fruits.
- Fasciculatum Group – red cone peppers, which grow erectly in clusters to a size of around 8 cm (3 in).
- Cerasiforme Group – cherry peppers, which are small and are fairly pungent.
- Conioides Group – cone peppers, which grow erectly and are very pungent.
- Longum Group – these peppers are pendulous when growing and taper to a point; they are often very mild.

Tabasco

Bell pepper

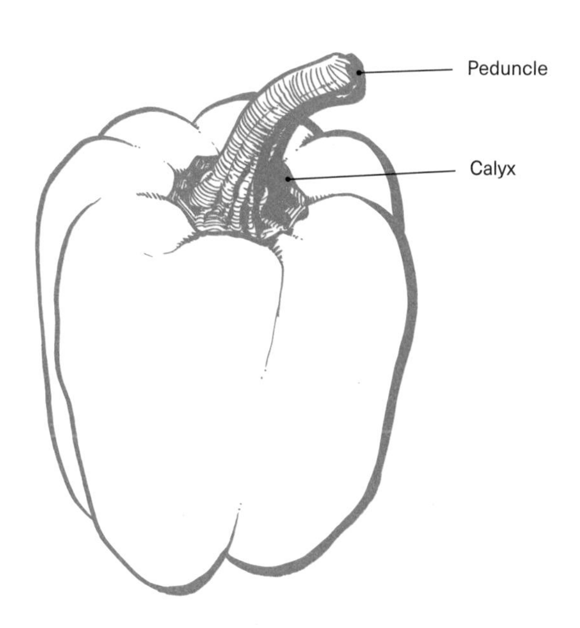

There is one rogue member of *C. annuum*, which at one point was its own species, but now is *C. annuum* var. *baccatum*. It looks very similar to the straight species, but the fruits are more globose and there are brown markings on the corolla.

Capsicum frutescens
These plants are short-lived perennials if they are not affected by frost, and famously host the pepper, tabasco. The flowers, which grow in the leaf axils in clusters of two or more, have blue anthers and a white corolla. The fruits produced are small, conical and with a fairly thin flesh, but are very pungent.

Capsicum chinense
These peppers are very similar to those produced by *C. frutescens*, except for a slight restriction below the calyx on the fruits. They are markedly pungent and include peppers such as habanero and the scotch bonnet.

Capsicum pubescens
Rocoto-type peppers fall into this species, and tend to have much darker-coloured flesh than other types. *C. pubescens* has a purple corolla and produces black, rugose seeds. Unlike the other species of *Capsicum*, it cannot cross-pollinate inter-specifically.

Cichorium intybus

Asteraceae

Chicory, Radicchio, Witloof

Today, *Cichorium intybus* or chicory is primarily cultivated for its edible leaves, but it has also been grown for its tap root. Chicory is renowned for a slightly bitter taste, caused by the metabolite sesquiterpene lactone, which is a part of the latex sap that runs through many Asteraceae plants and deters animals and insects from eating them. In its role as a plant defender, the

Radicchio

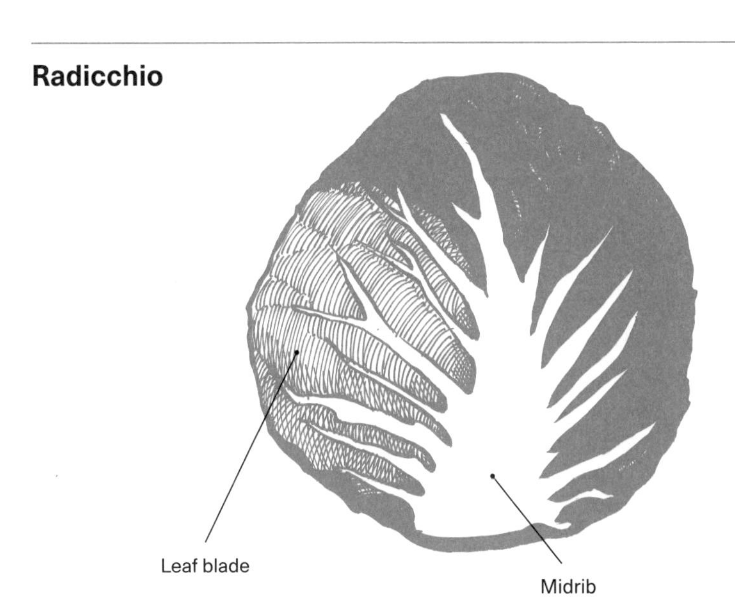

Chicory

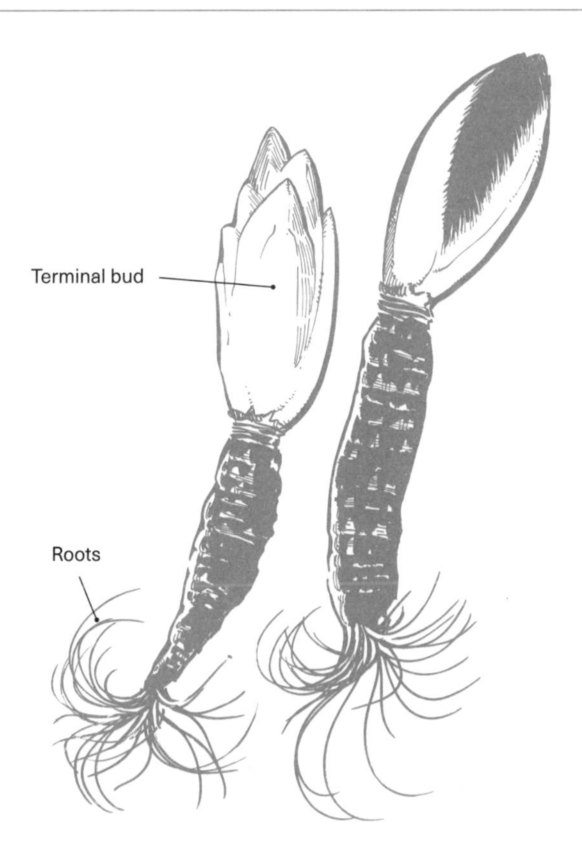

The botany of grafting and rootstocks

In the kitchen garden it is common to come across crops that are grafted onto rootstocks. Grafting is the process of fusing two different living tissues together to form one new plant. The bottom part of this plant is known as the rootstock, and most of it lives underground. It contains all of the roots, and the bottom-most part of the stem, but is not allowed to produce leaves, flowers or new shoots. Attached to this is the scion material, or the part that gardeners allow to fruit and flower. It is generally attached at the bottom of the main stem, and where the rootstock and scion meet, called the union, is often slightly swollen.

Grafting is the process of fusing two different living tissues together to form one new plant.

The rootstock starts life as a full plant, but while it is still young most of the aerial growth is cut back to just a stump. The scion also comes from a fully grown plant, but in many cases is just a shoot removed from a parent plant, as long as it has a healthy growth point at the tip. Where the rootstock and scion are to be joined should be similar in diameter to ensure a strong and successful graft. The aim is to join up the two cambium layers, which should lie in a similar place if the diameters are the same.

The cambium layer is the cells between the vascular tissues, the xylem and phloem, which transport water and soluble nutrients. The cambium is meristematic and retains the ability to differentiate into other types of tissue including secondary vascular tissue when needed, which is the property used when grafting two plants. The cambium lies in a ring around the outer edge of the stem in dicots, and is scattered in monocots, meaning monocots are extremely tricky to graft.

Once the rootstock and scion are placed together by a human, and wrapped tightly, the cambium multiplies and initially forms a callus material to heal the wound. This callus then differentiates further to form secondary phloem and xylem and link up the two plants.

Plants are grafted together to obtain the beneficial properties from each half. Quite often the rootstock can pass on disease resistance, the ability to withstand the cold or even make the scion flower

The most common reason for grafting is controlling the height of fruit trees. Apple trees, without rootstocks, would reach at least 7 m (25 ft), making them too tall for harvesting the fruit.

earlier. The most common reason for grafting is controlling the height of fruit trees. Apple trees, without rootstocks, would reach at least 7m (25ft), making them too tall for harvesting the fruit, so they are grafted onto rootstocks that control the eventual height, known as dwarfing rootstocks. The rootstock and the scion must be closely related, so apples are generally grafted onto other apples, and pears onto quince, which are both in the Rosaceae family.

Members of the Solanaceae family can also be grafted together, which is done in the first few weeks after the seedlings emerge. In this case, the rootstock is likely to be passing on resistance to diseases, such as blight in tomatoes.

bitter flavour tends to increase when the plant is stressed or near to producing seeds.

The leaves of chicory vary from green and speckled to deep reds, with contrasting white midribs and frilly green leaves. The red-leaved chicories are collectively known as radicchio. Most chicory leaves are hairy and stalked, growing in a rosette around a central point.

Chicory is a very hardy perennial in the wild, but often grown as a biennial in cultivation where, after a period of vernalisation and a day length of 13 hours, it will put up a flower stalk of around a metre, which produces a blue capitulum inflorescence. As with all plants in Asteraceae, chicory has a composite inflorescence made up of lots of smaller flowers, usually disc florets in the centre and ray florets on the outside. Chicory only has ray florets, which have a single, flattened corolla, and look like a single flower. In *C. intybus*, need other plants to cross-pollinated for successful seed setting. The resulting seeds have a pappus attached, which is a modified calyx and forms a parachute, which means the seeds can be dispersed on the wind.

Some chicory plants, usually known as Witloof, are used to grow chicons. This process occurs when the plant is dug up from the ground and the foliage cut back. The plant is then sunk into soil and kept in the dark at a temperature of around 18°C (65°F). This causes the terminal bud to grow and expand, growing leaves that do not contain chlorophyll and develop in a tightly packed formation, with a tendency to be less bitter.

Chicory grown for its tap root is classed as *C. intybus* var. *sativum*. In the past, the root was predominantly harvested, roasted and used as a caffeine-free coffee substitute.

Cichorium endivia

Endive, Escarole

Cichorium endivia is very easily mistaken for *C. intybus*, but there are some botanical differences to be spotted, the easiest being the lack of hair on the leaves of endive and the branched tap root underground. Endive is also an annual, not perennial, and can self-fertilise. *C. endivia* is either classed as endive, which has characteristic curled, lobed leaves, or escarole, which has broader, flatter leaves.

Citrus spp.

Rutaceae

Citrus fruits

The genus *Citrus* covers many similar edible fruits, collectively known as citrus fruit, with some of the most widely cultivated being oranges and lemons. Most likely originating from China and south-east Asia, they are all evergreen plants, meaning that they do not send carbohydrates down to their roots over winter and lose all their leaves, instead retaining their waxy, water-retaining leaves throughout the colder months. *Citrus* plants tend to have thorns, which appear in their leaf nodes and are made from modified cells from the stem. These thorns protect against herbivore attack.

The flowers that grow on citrus fruits are perfect, or hermaphrodite, containing both male and female parts, and they will self-pollinate. Some citrus fruits are also parthenocarpic, a trait desired by growers as they produce seedless fruits, which are much easier to sell to the consumer. Citrus also have a tendency to be polyembryonic, producing twins from a single fertilisation.

The fruits produced by *Citrus* spp. are entirely fleshy, so botanically they are berries, but have their own further classification of hesperidium, meaning fruits that have pulp in sections and a skin that can be separated off. The fruit is contained within a rind made of two distinct layers. First is the outside exocarp, known as the flavedo, which is pigmented and contains

Orange

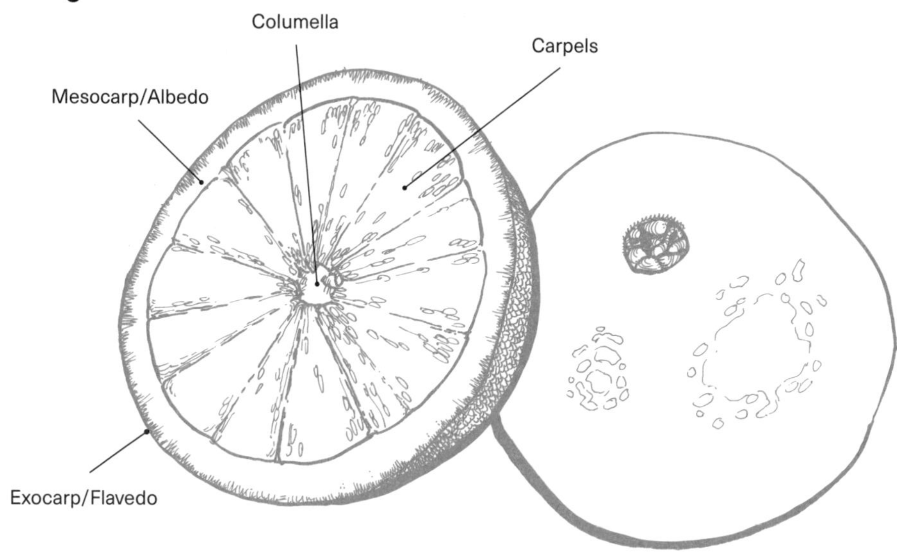

flavonoid compounds. The inner rind or mesocarp is known as the albedo and is white and spongy. Although the rind is edible, it is often discarded.

Within the rind are the pulp-containing carpels, which are arranged around a columella that is spongy like the albedo. The carpels or segments are contained within an endocarp membrane, and if produced via pollination will contain seeds. The segments contain a mass of multicellular trichomes, or hairs, which are hollow and filled with juice, and are known as the pulp vesicles. These spongy units protect the seeds from damage.

Citrus spp. mutate and hybridise rapidly in the wild, and many of the cultivated species are naturally occurring hybrids. A hybrid is when a distinct new species is created from the crossing of two parent plants, usually from different species. Hybrids can also be intergeneric, but these are much rarer.

Citrus* x *sinensis

Sweet orange, Navel orange, Blood orange

These evergreen trees can grow to around 15 m (50 ft) and are likely to have originated in south China. They are categorised by their relatively sweet flesh and a rind that is tight to the carpel segments.

The blood orange contains anthocyanin pigments in the pulp, which leads to red colours in the segments.

The navel orange is so called as on one end is a small opening that resembles a human navel. This opening is caused by the growth of a secondary set of carpels, which do not fully develop and fail to fuse properly. Each segment in this secondary ring is surrounded by a membrane, so is distinct. The rind that develops around the primary carpels grows up and around the secondary ring, but doesn't quite close, forming the navel.

Citrus* x *reticulata

Mandarin, Tangerine, Satsuma

This tree grows to around 8 m (26 ft) and is one of the hardiest of the *Citrus* spp. plants. It bears smaller fruits than *C.* x *sinensis*. The fruits have a rough rind that is not tight to the carpels and hangs a little looser, which makes it easier to peel off. Mandarins tend towards a yellower flesh, whereas tangerines are a deep yellow.

Citrus* x *aurantium

Bitter orange, Seville orange

Resembling *C.* x *sinensis*, the taste of these oranges is much more bitter owing to the presence of the compound neohesperidin. This makes the fruits fairly inedible raw, but suitable in conserves and marmalades, and as alcohol flavourings.

Citrus* x *limon

Lemon

This yellow-skinned fruit grows on a tree around 5 m (16 ft) tall. It has a sharp taste from the presence of citric acid.

Citrus* x *aurantifolia

Lime

A small, branched shrub 5 m (16 ft) tall, the lime has smaller fruits than the lemon, and is the most frost-sensitive of citrus plants.

Corylus avellana

Betulaceae

Hazelnut, Filbert, Cobnut

A European native, *Corylus avellana* or the hazelnut is a deciduous tree of around 6 m (20 ft), which often finds a home in the kitchen garden. As well as producing delicious, edible nuts, the wood is used as supports for climbing crops such as beans and sweet peas. The trees are coppiced, a process of cutting back the stems right down to the ground every three years – either the entire plant or a third of stems every year. The hazelnut coppices well as it tends towards basitony when apical dominance is removed via the coppicing process. In basitony, the lowest branches grow

Hazelnut

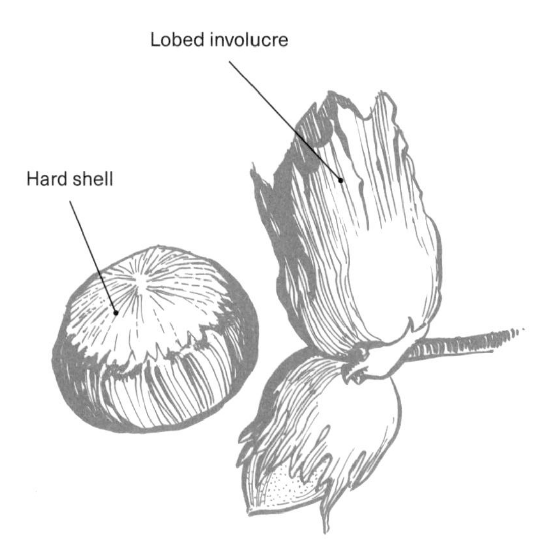

the strongest, producing long, straight poles for the kitchen garden.

The tree itself is monoecious, producing both male and female flowers separately but on the same plant. It cannot pollinate itself, and is entirely anemophilous or wind-pollinated, so it is a good plan to have several plants in the same area if growing for nut crops.

The male flowers hang in catkins, pendulous spikes of 240 simple flowers, which send their pollen into the wind to meet the female flowers. The catkins tend to be large and showy, with a yellow hue, and as they appear in winter when there are no leaves, they are very eye-catching. The female flowers take a little more hunting for as they are so small, but have startling crimson styles that appear at the end of the bud. Each bud contains several flowers, which are held in three bracteoles. After pollination, these turn into the lobed involucre or husk , which holds between one and four nuts.

The fruit produced by the hazelnut is botanically a true nut, which is defined as a large, hard fruit, that contains a single seed that does not split open when mature. The part eaten by humans is the nut held within the brown, hard shell. They can be harvested while still immature and green, and at this stage are known as cobnuts.

As well as *C. avellana*, other hazelnuts are often grown, including *C. maxima*, the fruits of which are commonly known as filberts. The nuts are larger than those of *C. avellana* and there are often six in the cluster.

Cucumis sativus

Cucurbitaceae

Cucumber, Gherkin

Cucumis sativus is a trailing vine that bears the fruit known as the cucumber, which although botanically a fruit, a seed-bearing organ formed

Cucumber

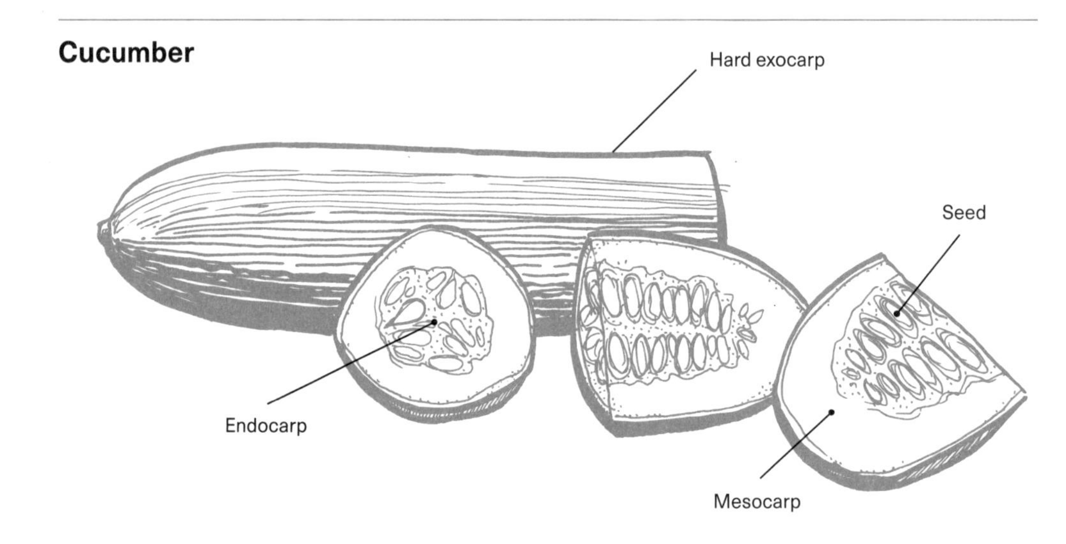

Cantaloupe

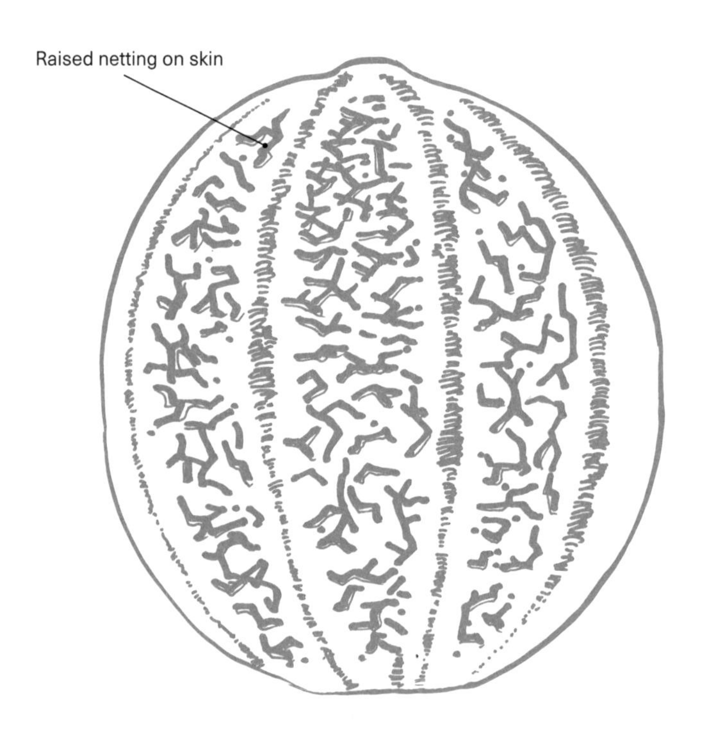

from the ovary, is eaten most commonly as a salad vegetable either fresh or pickled. The vine can trail or climb for up to 2 m (6.5 ft) and uses modified leaves that occur in the leaf axils and are known as tendrils to grab on to competitor plants to smoother them or climb up them in search of as much sunlight as possible. The cucumber is not hardy and will not survive a frost, and even struggles to thrive below 10°C (50°F).

The yellow flowers on the vines are day-neutral so can blossom at any time as long as it is warm enough. Cucumbers are generally monoecious, bearing both male and female flowers on the same plant, with the males appearing in clusters whereas the females are borne singly. Cucumbers will produce fruits via parthenocarpy where fertilisation does not take place and the resulting fruits have no seeds. These types are generally more desirable because pollination, which occurs via insects and mainly bees, causes the fruits to be bitter owing to the presence of cucurbitacin, a compound that is produced to deter herbivores.

Some types of cucumber are gynoecious, where they have predominantly female flowers with the occasional male for pollination, and others can be gynomonoecious, again with the majority of flowers being female but with some complete or hermaphroditic.

The resulting fruits are classified as a pepo berries, having a hard outer rind with fleshy insides that are not divided into segments. They generally have a green skin and are an oblong shape, although the cultivar 'Crystal Lemon' is yellow and round, so there is a large amount of variation. The whole vine is hairy, the hairs being formed of trichome, which protects the plant from predators, but the trichomes on the fruit are often much harder and are known as spines. The spines often combine with tubercles or warts on the skin to give a bumpy appearance. The flesh can be slightly tough but holds its shape when pickled.

Most cucumbers sold in supermarkets are known as burpless types. Many people found that cucumbers made them burp, thought to be caused by cucurbitacin, which is found predominantly in the seeds and skins. Burpless cucumbers are produced via parthenocarpy so have no seeds and have been bred to have much thinner skins, which reduces the burping effect.

Cucumis melo

Melon

The cousin of the cucumber is the melon, although scientists have put it into its own genus, *Melo*, at several points in history. The vine of *C. melo* is very similar to that of *C. sativus* but the fruits are much larger, more varied in colour and generally tend to be sweet so are eaten as a fruit.

There are several sub-categories of melon. These include cantaloupes, with a warty, grooved skin and aromatic flesh; musk melons, with raised 'netting' on their yellow or green skin and aromatic salmon flesh; and winter melons, whose green flesh is not aromatic and is surrounded by a smooth skin. The netting or raised areas on some melon skins are caused by fissures in the growth of the rind, which are healed by cells with suberised cell walls, basically forming a scab. It is thought this netting may add strength and protection to the fruit.

One fruit commonly called a melon, the water melon, is not actually *Cucumis melo*, but *Citrullus lanatus*; and one cucumber, the Armenian cucumber, which is not sweet and looks like a cucumber, is in fact *C. melo* var. *flexuosus*. Quite the confusing family.

***Cucurbita* spp.**

Cucurbitaceae

Squash, Pumpkin, Marrow, Courgette

As a group, these plants are commonly known by their generic epiphet of cucurbits. Originally from South America, in the wild they find themselves as either mesophytic (having adapted to grow in moist conditions) or xerophytic (thriving in areas with no available water). All *Cucurbita* species cultivated and grown in the modern kitchen garden are mesophytic, so require plenty of moisture.

A varied group of plants, cucurbits are grown for their fruits, which are eaten as vegetables, and include kitchen garden staples such as courgettes, squash and pumpkins. *Cucurbita* species are annual trailing vines, which will climb and scramble over other plants, forming adventitious roots at the nodes when they touch the ground. There are some forms that have shorter internodes and grow in a bushy fashion – in fact this is an easy way to tell if it is a summer squash, with short internodes, or a winter storing squash, which trails.

Squashes are annual, warm-season plants, which thrive at temperatures around 20–27°C (68–80°F). Their seeds will only germinate above 13.5°C (56°F). The leaves on squash are large and tend to be slightly lobed, some with silver mottling along the veins, which is often mistaken for the disease powdery mildew, which can plague cucurbits in hot, dry summers.

The plants are monoecious, producing separate male and female flowers on the same vine. The flowers are large and yellow, grow in the leaf axils and are pollinated by insects, mainly bees. The female flowers are easily identifiable by a large ovary growing just below the petals and are stimulated to grow by cooler temperatures and high light levels. The male flowers are produced when temperatures are higher, which means some summer plants may predominantly produce male flowers and the vines will not yield many fruits. The flowers

Pumpkin

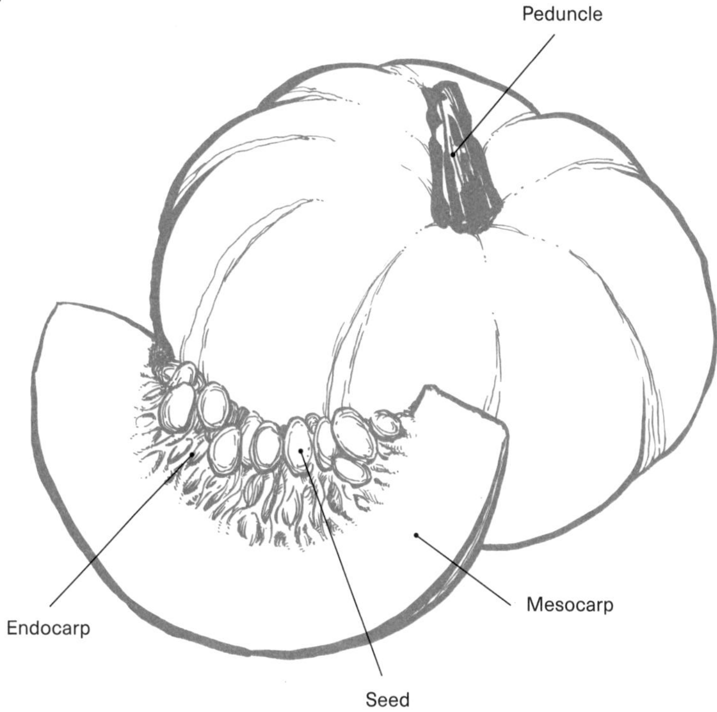

are edible, and many chefs enjoy stuffing the large yellow blooms, but growers will attempt to harvest only male flowers, those without the small bump of an ovary at their base.

Once pollinated, the ovary swells, and the petals fall off, eventually leaving a dark scar on the bottom of the fruits. The hypanthium or receptacle tissue fuses with the ovary to form the exocarp or rind, which surrounds the firm mesocarp flesh that is mainly what is eaten by humans. In the centre of the fruit is the very soft endocarp, which is generally scooped out with the seeds, along with placental tissue that connects the seeds to the fruit. The peduncle lignifies, getting woody, and forms a convenient handle at the top of the fruit.

The fruit itself is classified as pepo, with a hard rind and soft insides. The rind in some *Cucurbita* spp. lignifies and becomes quite hard, especially ornamental types, but owing to a recessive allele, most of the squash we consume do not lignify in the rind and can be cut into easily.

The fruits are left to ripen on the vine for as long possible, and when harvested contain a large amount of stored starch. They are then often 'cured' post-harvest. This process keeps them in a warm environment for a few weeks during which the cells break down the starch to release sugars for energy, which makes the fruits sweeter. This process also often hardens up the rind and allows them to store for longer.

There are many species of *Cucurbita* commonly grown in the kitchen garden, but the three most common forms are covered below.

Cucurbita pepo

Courgette, Marrow, Patty pan, Summer squash, Pumpkin, Acorn squash, Delicata squash, Spaghetti squash

Cucurbita pepo is probably the most widely grown of the *Cucurbita* species and produces both summer and winter squash. The summer squash types include courgettes, marrows and patty pans. These plants mainly grow as bushes, with a few climbing exceptions, and the fruits are harvested at an immature stage when the skin is still very soft. If left to mature, the fruits of these cultivars would get harder with lignin but the interior would hollow out and the flesh would become stringy and not particularly palatable. Summer squash all have white flesh contained within skin that varies in colour from white to green and yellow.

Winter squash have orange or yellow flesh and a skin that can be orange, yellow, green, blue or even white. The fruits are highly polymorphic, coming in a range of shapes, from the traditional pumpkin shape, round and orange, to oblong types such as delicata. The stems are fairly rigid with grooves and spiky hairs, which are also found on the deeply lobed leaves and protect against damage from herbivores. The peduncles are star-shaped in cross-section and, unlike other species, do not flare where they attach to the fruit.

Cucurbita moschata

Butternut squash, Crookneck squash

These winter squash can be ribbed or smooth-skinned and are generally a pale red or brown with a yellow or orange flesh. Most commonly grown are the butternut types, with a slim neck and fatter base. These squash thrive in hotter conditions than most, which can mean they are hard to ripen in cooler climates. They often have silver mottling on their leaves and feature a smooth, pentagonal peduncle, which flares at the base.

Cucurbita maxima

Pumpkin, Buttercup squash, Hubbard squash

As the specific epithet *maxima* suggests, the fruits of these plants tend to be larger than the other species, and most giant pumpkins are *C. maxima.* The fruits are generally round and yellow or orange and store for longer than other species. The stems are soft and round, while the peduncle is spongy inside and cylindrical in cross-section.

Cynara scolymus

Asteraceae

Globe artichoke

Originating in the Mediterranean, this herbaceous perennial plant goes by the common name of globe artichoke or just artichoke. Cultivated in the kitchen garden for its edible, immature flower buds, artichokes are also often seen in ornamental borders thanks to their extremely attractive grey–green leaves, which are highly dissected, and violet flower heads. They grow well at temperatures around 12–18°C (53–64°F), but oncethey reach 22°C (72°F), the buds become tougher, and unpleasant to eat.

Being perennial, the artichoke has a large fleshy tap root in which it stores reserves over winter to allow it to grow again in the spring. It is cross-pollinated by insects, particularly bees, but very rarely produces viable seeds, and those that are produced are so variable that the artichoke is mainly propagated vegetatively. Using stumps, or sections of root with basal stems attached, the root section provides energy for the basal stem to

Artichoke

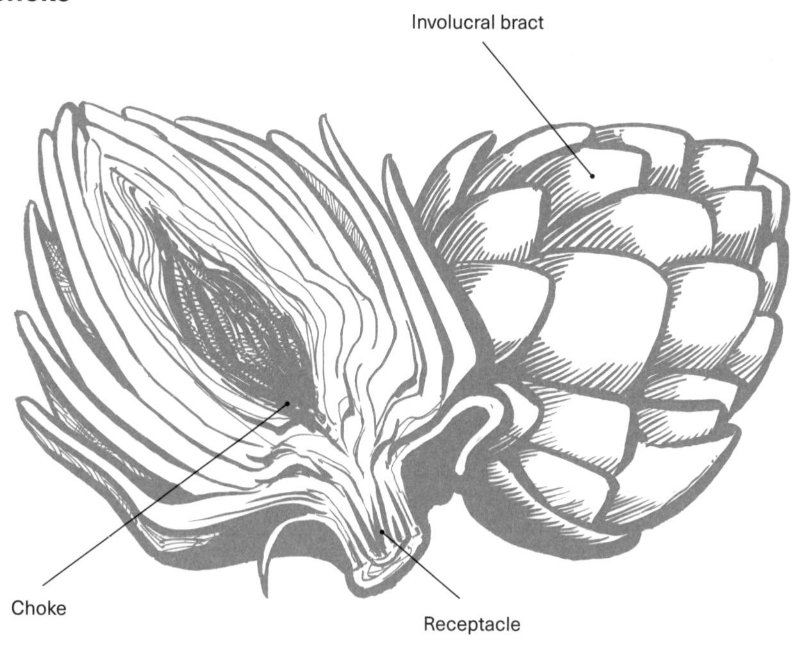

grow up and put out initial roots to form a clone of the parent.

The edible portion of the plant is the giant globe flower head, which, as it is in the family *Asteraceae* and has the typical capitulum inflorescence, is actually hundreds of flower buds, which will eventually turn into blue tubular disc florets. Although it is often thought that it is the flower buds that are eaten, these immature flowers are discarded once harvested, as they form the 'choke' from which this crop gets its name. They are in the very centre of the head and each one contains a tiny receptacular bract, known as chaff or palea, which is highly irritating if eaten, and therefore entirely removed. The heart of the artichoke, which is the most tender, is in fact the young receptacle, the end of the stem where all the flower parts are attached.

As well as the heart, the large, fleshy, outer bracts are also eaten. Bracts are modified leaves and are present in many plants, often highly coloured and resembling petals. In the artichoke, the bracts act as protection for the inflorescence, and the outer ones have spines to stop herbivores eating the plant. These outer bracts are called the involucral bracts or phyllaries and are the main part of the harvest. The outer ones are the most fleshy, but also most tough, and towards the centre they become pale and more tender where they have not produced chlorophyll as they are not exposed to the sun.

The botany of pruning

Pruning plants, particularly woody ones such as trees and shrubs, is generally undertaken to control their size and shape, but also to promote the best crop production possible. Pruning artificially manipulates the hormone flow around the plant to give the desired outcome. Hormones are often produced in one area, and transported around to where they are required, and pruning particularly disrupts the auxin–cytokinin pathway.

Pruning artificially manipulates the hormone flow around the plant to give the desired outcome.

Auxin is produced in the apical buds of limbs and flows downwards to promote root and shoot growth. Auxin is also responsible for cell elongation, and fruit drop or retention, but in the pruning process it is its involvement in apical dominance that is important. Apical dominance is where the terminal bud is the only one that grows upwards, and axillary buds remain dormant. In the wild, this is an important feature as plants need to reach upwards for the light and often need to outcompete other plants to become the tallest and not live in the shade of others. Apical dominance is achieved via auxin's interaction with cytokinins.

Cytokinins are produced in the root tips and flow up the plant to promote canopy and bud growth. They are involved in cell division and will influence cell differentiation when needed. It is the bud growth that is the main property involved in pruning.

The concentration of cytokinin is much weaker than that of auxin, which means that when they meet, it is the auxin that wins, and the axillary buds remain dormant. Once the apical bud is removed and auxin is no longer produced, the cytokinin can go to work, promoting the growth of the previously dormant axillary buds. The topmost bud will receive most of the hormones, and often after time become the leader, so its apical bud will once again start producing auxin and inhibiting the growth of the axillary buds below. When removing an entire limb, it is the epicormic buds that are promoted to grow, these being located under the bark and invisible until they are required.

When pruning takes place when the plant is in full growth, this is generally called restrictive pruning, whereas dormant pruning is referred to as unrestricted.

Another important aspect of pruning is timing. When pruning takes place when the plant is in full growth, this is generally called restrictive pruning, whereas dormant pruning is referred to as unrestricted. What is being affected here is the volume of the tree in comparison to the volume of stored carbohydrates and nutrients over the dormant period in the roots.

If a plant is pruned while it is in growth, by the time is goes dormant, the volume of stores is the same as the tree, so when spring occurs and the stores flow through the tree again, there is the right amount and the growth is steady and controlled. If the tree is pruned while dormant, the volume of the tree decreases and becomes less than that of the stores. So when spring arrives and the stores flow through the tree, the excess amounts produce lots of new growth, especially breaking new buds and creating lots of young stems.

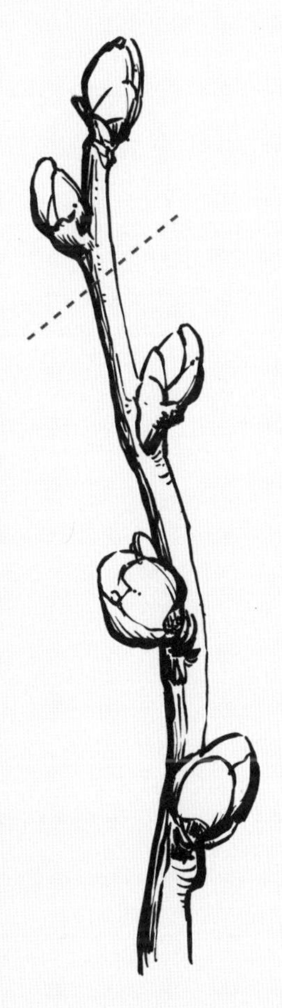

Daucus carota

Apiaceae

Carrot

This delicious umbellifer is grown for its edible tap root, although its leaves are also a tasty treat. The root is the main storage organ of the carrot, and therefore the part that has been cultivated by humans, from a small, white, scrawny harvest in the wild, to the large, fleshy orange delight grown in kitchen gardens today. The root is a fleshy, conical tap root, with the upper portion made of the swollen hypocotyl, which then produces the tap root, which swells to become the lower part of the carrot.

Within the root crop, there are two main noticeable sections. The centre is the core, made up of xylem, and surrounding this is the cortex, which is the phloem. Carrots have been bred to have as much cortex as possible, as the core tends to be tougher in texture. When growing, the carrot will put down its full length first, and then begin to build up girth, quickly at the top and more slowly at the bottom. The top of the carrot is known as the shoulders, and these can be narrow, or extremely wide in varieties such as the Oxheart. On the outside are lots of scars where lateral roots were growing that have since fallen away after harvest.

Although the average carrot is orange, this has not always been the case. Originally white, it is thought a sudden mutation in Afghanistan led to the existence of purple carrots, starting with a white inner and purple skin, then being bred to a fully purple carrot. These eastern carrots contain anthocyanin pigments, and the roots frequently branch. Along with the eastern carrot, there is another distinct cultigen, the western carrot. It is likely that breeders were working on the wild white carrot when a yellow-fleshed carrot was created. Later, in the Netherlands in the 16th and 17th centuries, this yellow tone was deepened during the breeding process to create the sweet, orange carrot sold worldwide today.

The orange pigment in carrots comes from carotene, which creates vitamin A when eaten by humans. Carotenoids play an important protective role in plant cells, especially during photosynthesis where they can absorb any excess light energy collected, so it does not damage the cell. Carotenoids can also increase the amount of photosynthesis a cell can undertake, as they absorb light at a different wavelength from chlorophyll, increasing the range the cell can access.

Carrots are also notorious for forking, or splitting into two tap roots. The root itself grows down through the soil and is protected in this movement by the root cap, which is constantly replaced as it makes this journey. The root cap also produces a mucus-type substance to help lubricate the downwards growth. If damaged severely, the next points of growth of the root are two secondary tap roots, which split. So if the root hits a large stone or similar, it will fork.

Eruca vesicaria

Brassicaceae

Rocket, Arugula

A native of the Mediterranean, the annual crop *Eruca vesicaria* is more commonly referred to as rocket or arugula, and is grown predominantly for its leaves. The type cultivated in the kitchen garden is *E. vesicaria* subsp. *sativa*, and as it is in the family *Brassicaceae,* the leaves contain the mustard oils that these plants produce as a defence against herbivores. The leaves are dark green, and pinnate with deep lobes, terminating in one large lobe, looking very appetising in a salad.

The plant grows to around 60 cm (23 in) and produces cream flowers with red veining. These are hermaphroditic or complete, containing both

Carrot

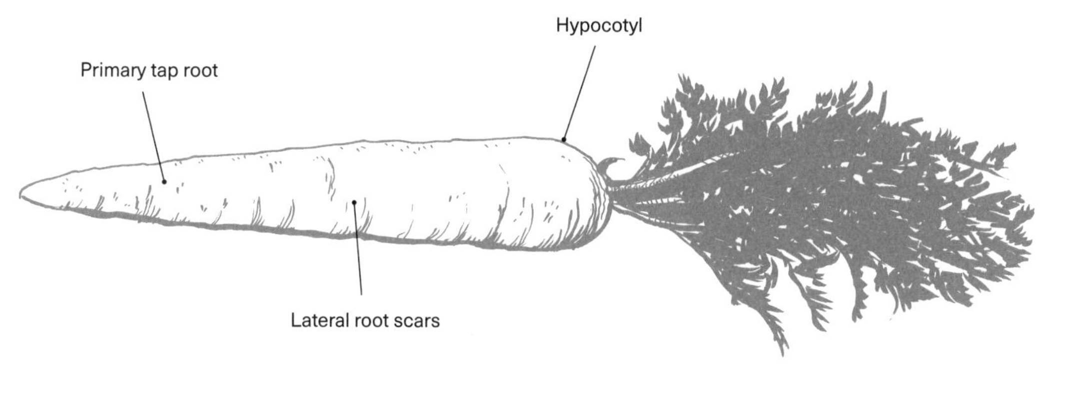

Rocket

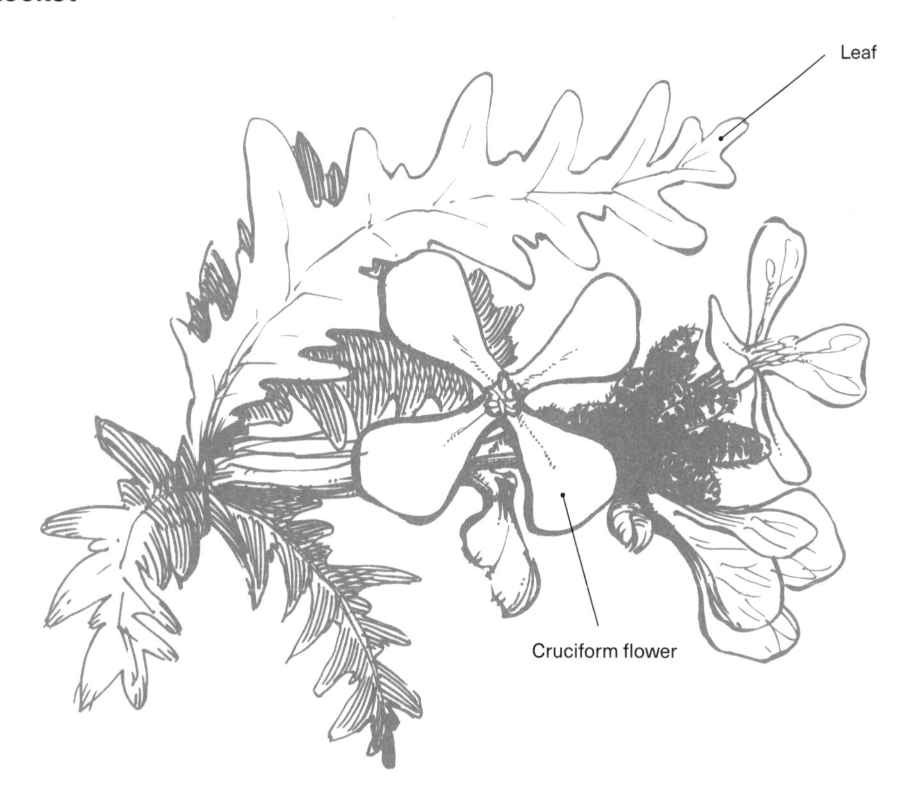

male and female parts. This is a long-day plant and is triggered to produce seed in the summer months, which means it has a tendency to bolt in the hotter months. Bolting is when a plant produces a flower spike and seeds prematurely, usually when there is stress from temperature or attack, in the hope it can produce the next generation before dying.

Confusingly, there are two plants referred to as rocket, both in the Brassicaceae family and both cultivated for their leaves. The second is *Diplotaxis tenuifolia*, also hailing from the Mediterranean, and also referred to as wild rocket and perennial wall rocket. Although very similar, this is a good example of where knowing the Latin name of a plant can help understand it. Many features are shared by *D. tenuifolia* and *E. vesicaria*: they grow to similar sizes, have pinnate leaves and taste very similar. *D. tenuifolia* has yellow flowers rather than cream, and another important difference is that it is a perennial, whereas *E. vesicaria* is annual

As *E. vesicaria* is annual, it puts more energy into producing seeds, making larger ones than those of *D. tenuifolia*. The seed produced by the annual will germinate more readily and at a larger range of temperatures, as this is *E. vesicaria*'s only way of continuing the line; it also grows faster than *D. tenuifolia*. As a perennial, *D. tenuifolia* will gain size more slowly, because it can draw on stored energy and doesn't need leaves to photosynthesise for its first hit of energy in the spring, and it will also grow back if all the leaves are harvested, making it a useful 'cut-and-come-again' plant.

Ficus carica

Moraceae

Fig

The fig is a large, deciduous tree, which can grow anywhere from 2 to 5 metres (6–16ft) in height. Originating in Asia, it is adapted to live in drought conditions, but has proven it can very happily make use of excess water, as long as the ground never becomes waterlogged. When wounded, the fig will produce a latex, which heals the wound but is also the reason gardeners tend to prune figs when dormant, to stop excessive loss of sap. Although coming from warm climes, the fig can live through colds of −9 to −7°C (15–20°F).

The fruit of the fig tree Is a curious one, coming in hues of green, purple and brown, and needing eight hours of daylight to ripen fully. The fruit starts life as a syconium, a hollow receptacle that will become the fruit, within which all the flowers are borne, so a gardener will never actually get to view the small inflorescence of the fig. This receptacle is the extended peduncle, which will eventually become the bulk of the sweet flesh of the crop. From the outside, the only other visible part of the fig is the ostiole, which is a small opening on the bottom that allows pollinators access and is defended by sharp bracts.

Within the syconium are the fig's flowers, which develop into an aggregate fruit. There is some discussion about whether the individual fruits are drupes or achenes, so either fleshy or dry, but as the seeds themselves are dry, and the fleshy part of the fruit is the pedicels that attach the seed to the syconium wall, achene is now the more generally accepted term.

Before becoming the fruit, the fig must undergo pollination, and this is where things get even more mysterious. The fig is gynodioecious, meaning that within the population there are female and hermaphrodite plants, and within *F. carica* there are several different types of tree with different flower production and pollination traits – caprifigs, Smyrna figs, common figs and San Pedro figs.

In the wild, figs are pollinated by a tiny wasp, a type of pollination called caprification. Caprifigs

Fig

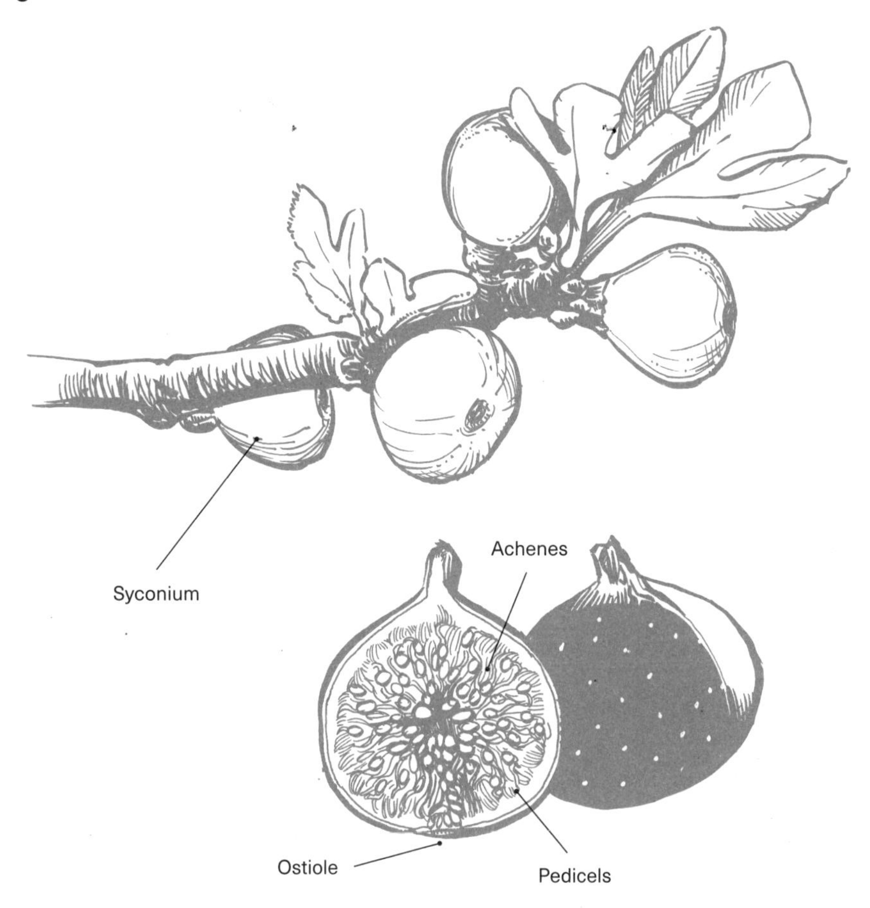

are pollinated this way, where the wasps hatch within the fruit in the spring, and the wingless males mate with the females, then help to chew a hole out via the ostiole, before dying. The females then leave this fig covered in pollen and chew their way into the female syconium. Upon entry, the wings of the wasp are torn off so she cannot leave, so she lays her eggs and dies here. Old knowledge claims that the crunchy bits of the fig are the wasp's body, but this is untrue as she is broken down by enzymes, and the crunch is simply the seed.

Caprifigs are the only fig to produce male flowers, known as profichi, and these only occur in spring. The male syconium also contains female flowers, but as these are short-styled flowers whose only function is to allow wasps to lay eggs there, the fruit is still classed as male as it primarily produces pollen, although it is in truth hermaphrodite. The profichi are inedible as a crop, but useful as pollinators, with growers often taking these figs and placing them around female trees so the wasps within are released close to the flowers that need pollinating.

Smyrna figs produce only female syconia, and need pollinating by the wasps emerging from the caprifig. The flowers within are only long-styled, which means they can be pollinated, but there are no short-styled flowers to act as a nursery for the wasp, therefore she cannot lay her eggs.

The common fig is the one mainly grown, and due to a mutation around 11,000 years ago, produces fruits through parthenocarpy, meaning no pollinator is needed, which is particularly useful in areas of the world where the tiny wasp does not live. The San Pedro figs also give only females, but can produce through both pollination and parthenocarpy.

Fig trees can be classified further as either uniferous or biferous, depending on how many cycles of fruit production they undergo in a season. Uniferous trees produce one main crop, on the previous season's growth, starting their life cycle in spring and being harvested in autumn. Biferous trees produce two crops in one year, the main crop and breba fruits. Breba are often formed at the very end of the season, in late autumn, so appear on the previous season's growth in the spring. They are often killed by extreme cold, so it is not common to have breba fruits in colder areas. The breba are harvested in early summer.

Common figs can produce both breba and main crop, both through parthenocarpy. San Pedro produce breba through parthenocarpy, but their main crops need pollination, and caprifigs can produce up to three cycles of fruit (they are 'triferous'). The Smyrna fig does give breba, but as the caprifig only really has male pollen in spring, these late figs are rarely pollinated and are aborted.

Foeniculum vulgare* var. *azoricum

Apiaceae

Florence fennel

A fragrant member of the kitchen garden, *Foeniculum vulgare* var. *azoricum* is commonly referred to as Florence fennel or bulb fennel, which distinguishes it from the tall, feathery fennel plants grown in the herb patch whose Latin name is *Foeniculum vulgare* var. *dulce.*

Florence fennel is native to the Mediterranean and as such grows well in warm conditions, although it is hardy enough to survive through a few frosts. It is a perennial plant but is grown as an annual and harvested for its white pseudobulb. Called a bulb of fennel, it is actually the swollen base of the stem that people pick and eat so is not technically a bulb, which is an underground storage organ. The fennel 'bulb' sits on top of the soil and is occasionally referred to as a fennel apple. It is a very flavoursome crop, tasting strongly of aniseed, but containing up to 87 different volatile compounds, predominantly used as pest deterrents by the plant.

The pseudobulb will grow to around 30 cm (12 in) and if not harvested will then continue its life cycle by producing a hollow stem that bears its inflorescence. Flowering in fennel is controlled by photoperiodism, where the plant senses periods of dark. Fennel is a long-day plant and day lengths of 13.5 hours will initiate flowering, so generally by midsummer fennel will flower, no matter how large the pseudobulb is, meaning it is best to harvest the crop by late spring, or put off sowing it until late summer.

Both bulb fennel and the herb fennel produce flowers arranged in an umbel formation, which is common in the *Apiaceae* family. An umbel is a cluster of small flowers on stalks or pedicels of roughly equal length that form a flat or domed platform of flower heads that make a perfect landing pad for insects to sit and feed from the

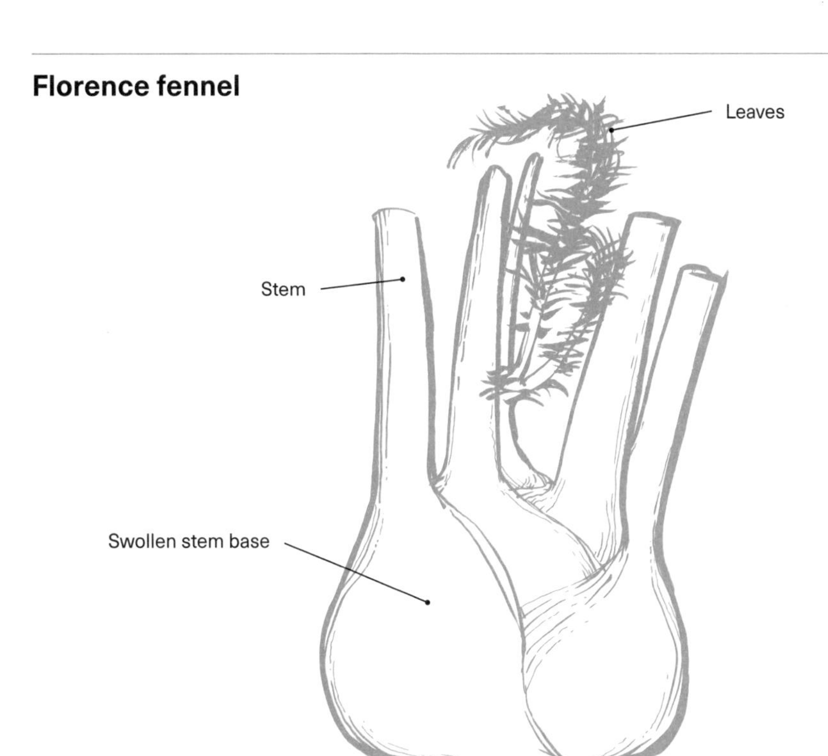

flowers. Fennel flowers are hermaphrodite, insect-pollinated and self-fertile. The resulting fruit and seeds are edible from both varieties of fennel. The fruit of fennel is a schizocarp, a dry fruit that splits open when mature. It is attached to the seed via its pericarp, so the fruit and seed are often referred to as the seed when harvested as a spice.

Fragaria x *ananassa*

Rosaceae

Strawberry

One of the popular summer fruits of the kitchen garden, *Fragaria* x *ananassa*, otherwise known as the strawberry, is a red, juicy delight on a hot day. It is a perennial, herbaceous plant, which dies back overwinter, storing energy in its crown and growing leaves each spring. As the 'x' in the Latin name shows, the common strawberry is in fact a hybrid of two other species of strawberry, in this case two American types – a woodland plant called *F. virginiana* and the evergreen *F. chiloensis*, which has extremely large berries. Although they were native American types, the cross actually occurred in France in 1780.

Strawberries have white flowers, which are hermaphroditic and self-fertile, although the fruit produced tends to be better and larger when cross-pollination occurs. The plants are mainly insect-pollinated, although the wind will

The botany of plant nutrition

Crops will often show physical signs that they are lacking in certain nutrients, such as yellowing leaves or stunted growth.

Much time in the kitchen garden is spent ensuring that crops have all the nutrients they need to be strong and healthy. Crops will often show physical signs that they are lacking in certain nutrients, such as yellowing leaves or stunted growth. Although fertilisers can be added, a healthy soil will provide everything the plant needs. The three main nutrients required by plants are nitrogen (N), phosphorus (P) and potassium (K), which are the main constituents of inorganic fertilisers. The other macronutrients required by plants are magnesium (Mg), calcium (Ca) and sulphur (S).

Nitrogen is used by the plant to produce amino acids, which are the building blocks of proteins. One of the most crucial proteins it produces is chlorophyll, the green pigment that absorbs light energy in the process of photosynthesis. As chlorophyll is mainly stored in the green leaves of the plant, nitrogen is associated with vegetative growth and crops such as spinach and brassicas, although it is essential for all plants. Nitrogen is found within organic matter in the soil, where it is slowly decomposed into a form the plant can take up. Nitrogen fertilisers are already in a form the plant needs, but these molecules are also very mobile in the soil and are leached away with water, which is why the nitrogen slowly provided by composts and healthy soils is more valuable.

Also readily found in soils is **phosphorus**, although this needs to be converted by soil organisms to a form readily taken up by the plant. Phosphorus is used to create

the compound adenosine triphosphate (ATP), which is used to transfer energy in the cells. ATP is critically important in the meristems of the roots and shoots of a plant, especially in the seedling stage, to ensure healthy growth.

Potassium is used by the plant to act as an osmotic regulator, which means it controls the movement of water in the cells and is vital to healthy plant function. It is also used to resist chilling, injury and drought. Unlike nitrogen and phosphorus, the plant never converts the potassium ions it takes up into another compound. This nutrient is found in the clay particles of soil and is released by weathering.

Magnesium, calcium and sulphur are also required for healthy growth but all are readily found in soils so very rarely need to be artificially added.

Magnesium, **calcium** and **sulphur** are also required for healthy growth but all are readily found in soils so very rarely need to be artificially added. Magnesium and sulphur are major components of chlorophyll among other proteins, so are essential for plant growth. Calcium is used by the cells to fuse their walls, again making it a very important nutrient.

All of these macronutrients can be artificially added when required, but a healthy soil with organic matter containing the nutrients and a variety of soil organisms to release the nutrients will provide the crops with a steady food source. In smaller amounts, plants also need **trace elements**, which are generally readily found in soil. These include iron for chlorophyll synthesis, manganese, zinc and copper to form enzymes, boron for plant processes such as the movement of sugars, and molybdenum to assist with the uptake of nitrogen.

Strawberry

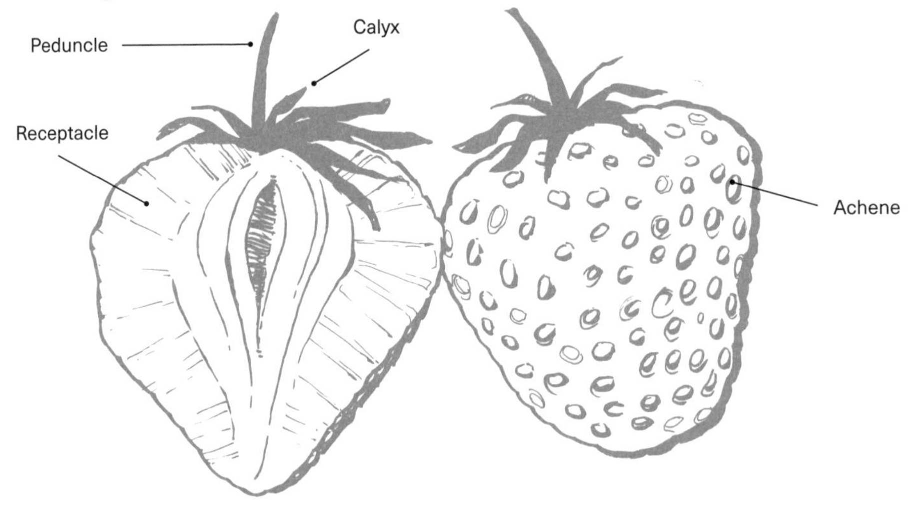

carry a little pollen. Once pollinated, the fruit is produced. Although known to most as a berry, this is not a botanically accurate description of the fruit of *F.* x *ananassa*, which is commonly known as a false fruit, pseudocarp or accessory fruit, as the majority of the flesh is not produced from the ovary.

In the case of the strawberry, most of the sweet flesh is the swollen receptacle, which has the fruits embedded in it. The receptacle is often hollow in the middle as the outside grows so much faster than the inside that it never quite fills up. The fruits embedded within the receptacle are of the achene type, single-seeded, dry and indehiscent. What are commonly referred to as seeds are the actual dry fruits that contain a seed within. They are formed from the 500 pistils that cover the receptacle in a spiral pattern.

If not all pistils are fertilised, the strawberry is often deformed or not very large. Deformed fruits are known as nubbins, or button berries, and some deformations look like several strawberries stuck together, known as cockscomb strawberries.

On the top of the fruit is the pedicel, which lies above a green fringe of sepals, known as the calyx. The epicalyx below this is formed of bracts, which are stipules that have fused together and help protect the fruit and flowers. Look closely and the sepals and bracts appear alternately. Occasionally found underneath the epicalyx are dried petals, but also the stamens, which sit on top of the fleshy receptacle.

The seeds produced are viable and spread via the digestive tracts of herbivores. But strawberries also reproduce vegetatively, making clones of the parent plant. They do this by putting out runners, or stolons. Stolons are modified stems that grow horizontally, and in the case of the strawberry along the top of the ground. At the nodes of the stem, the runner puts down adventitious roots, which anchor the stolon,

Jerusalem artichoke

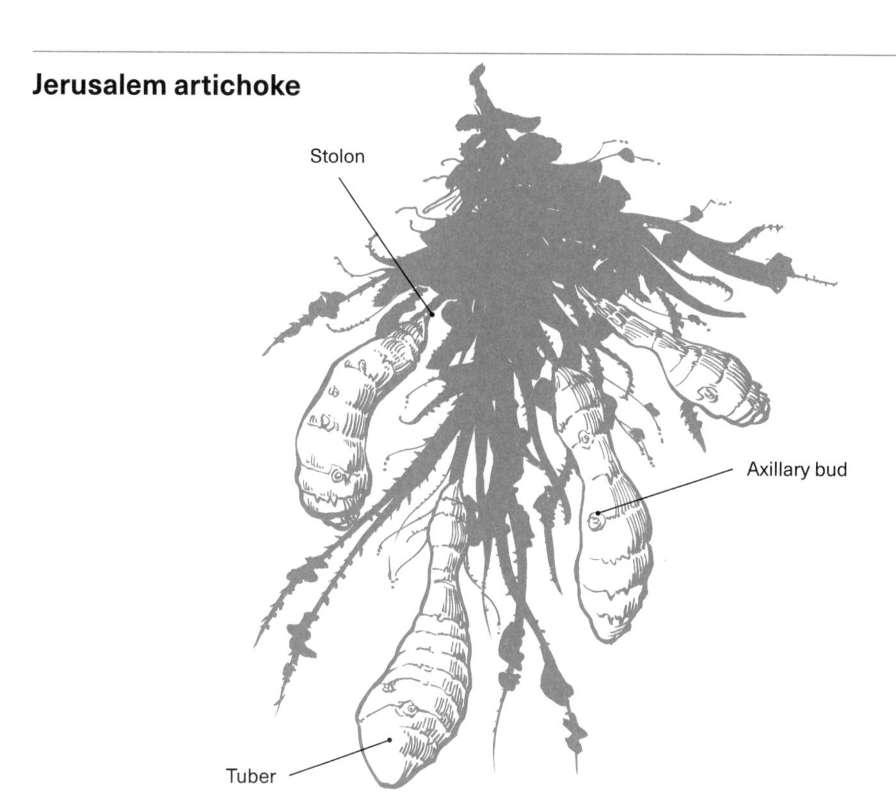

and then leaves are formed from the bud at the node. These daughter plants are identical to the parent, which is beneficial in evolutionary terms as it has proven to be thriving, and by sending the daughter plants out along a stolon, the strawberry can put down roots in the best environment close by and colonise the area.

Helianthus tuberosus

Asteraceae

Jerusalem artichoke

Helianthus tuberosus is a tall, perennial, herbaceous plant that is primarily cultivated for its edible tubers. Commonly known as Jerusalem artichoke, it neither originates in Jerusalem, nor is an artichoke, but instead is a relative of the sunflower – a fact that becomes obvious when the 3 m (10 ft) plants put out their sunny yellow flowers. The flowers only appear when the day length becomes shorter, at the end of the summer, at around 13 hours. The foliage dies back over winter, but the plant stores food in its tuber to help shoot again the following spring.

The tuber is primarily the harvest that *Helianthus tuberosus* is cultivated for, although the leaves make a good fodder crop. In the kitchen garden, the Jerusalem artichoke is often classified as a root crop, but is in fact a true tuber. It is a modified stem that originates from the axil of a lower leaf, then grows underground and stores carbohydrates. The tuber is connected to the stem via stolons, and on its surface are lots of lumps, which are the axillary buds that will burst through underground and produce new plants.

Unlike many storage organs, the Jerusalem artichoke stores its carbohydrates in the form of a sugar, and not starch. The sugar in question is called inulin, which makes the crop fairly sweet. Inulin can be processed to become fructose, so this tuberous crop may be a future commercial provider of sugar. This process happens naturally, albeit slowly, when the tubers are kept in storage and respire slightly.

The one drawback of inulin is that it is not easily digestible by humans, and it passes through the digestive tract untouched until it reaches the colon, where it breaks down and releases a gas, which can give a slightly unpleasant side effect to the consumer of this harvest.

Ipomoea batatas

Convolvulaceae

Sweet potato, Yam

No longer found growing in the wild, it is thought that *Ipomoea batatas* originated in tropical America, and its need for warm, humid, frost-free conditions supports this. Generally known as sweet potato, it is not related to the potato, *Solanum tuberosum*, and unlike the potato, which is a stem tuber, it is a true root crop as its underground tubers are swollen roots. In some countries, the sweet potato is known as yam, which again is a confusing moniker, as more often yam refers to the edible tubers of a plant called *Dioscorea* spp.

The sweet potato is a herbaceous vine that grows along the ground, with up to 5 m (16 ft) spread. However some types have shorter internodes, the gap between the leaves, giving bushier plants. Although commonly grown as an annual in the kitchen garden, left unharvested it

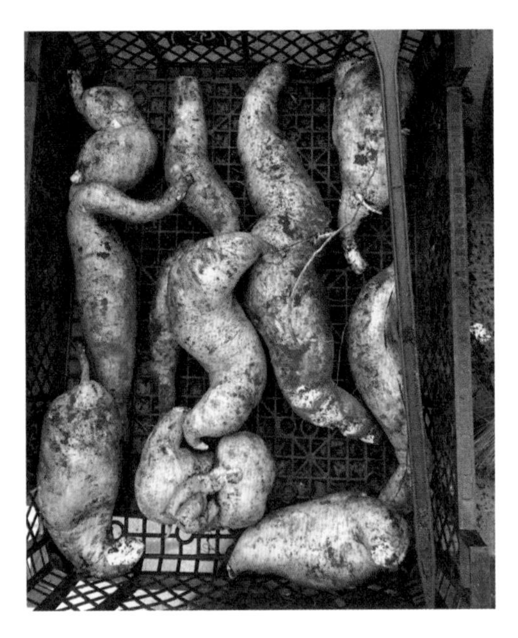

Sweet potato

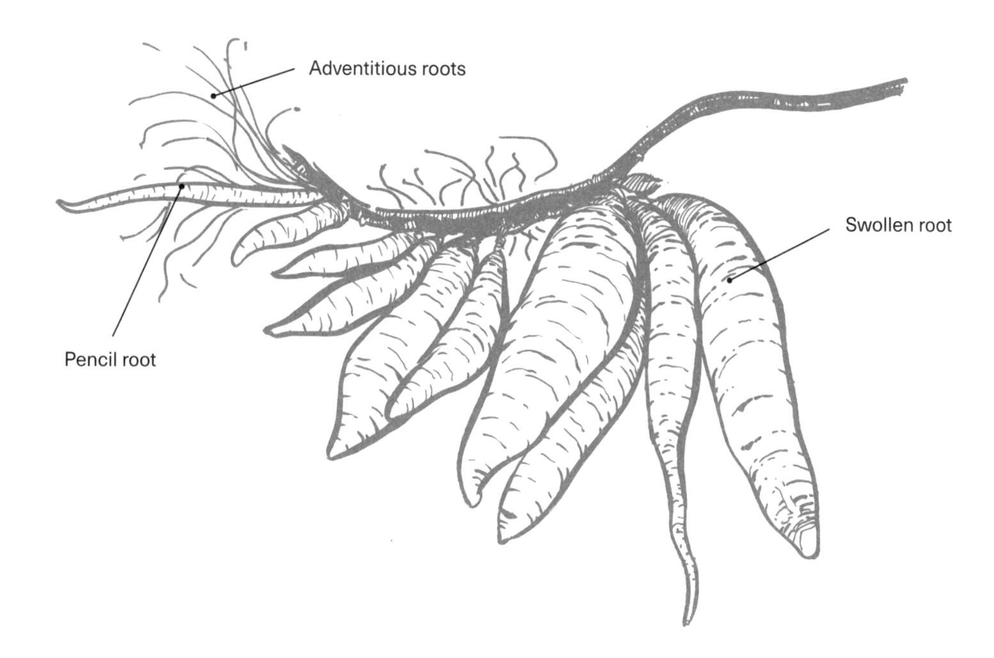

is a perennial plant, which is why it stores food reserves in its roots, to get it through colder seasons. The nodes along the stem, where the leaves form, will put down adventitious roots where they touch the ground, producing new plants. These roots are initially slim, to help travel through the earth, but once the plant starts to put on leaves, it can produce enough energy to store food.

Sweet potatoes start to form tubers when the day length reduces, and it is known as a short-day plant. The adventitious roots are able to modify into one of three types of root, namely fibrous, pencil and storage. Fibrous roots are extremely slim and can take up water from the soil, but cannot store food. Pencil roots are produced when conditions are not perfect, and they bulk up slightly, but stay relatively small, about the thickness of a pencil. Storage roots are the ones that bulk up to produce tubers, which humans harvest. This is not a true tuber, as they are formed from modified stems, but is known as a root tuber. It occurs when the root begins secondary thickening and the cambium cells between the xylem and phloem vessels differentiate to become storage parenchyma cells where the plant can place starch that it uses the next spring to respire.

The sweet potato itself comes in an array of colours. The flesh can be orange, red or white, and the skin varies from purple through to cream. On the skin are lines of parallel scars, which are where small roots grow, and if these are still attached, they will be growing downwards, towards the distal end, or the end that grows away from the surface. This can be important to

Lettuce

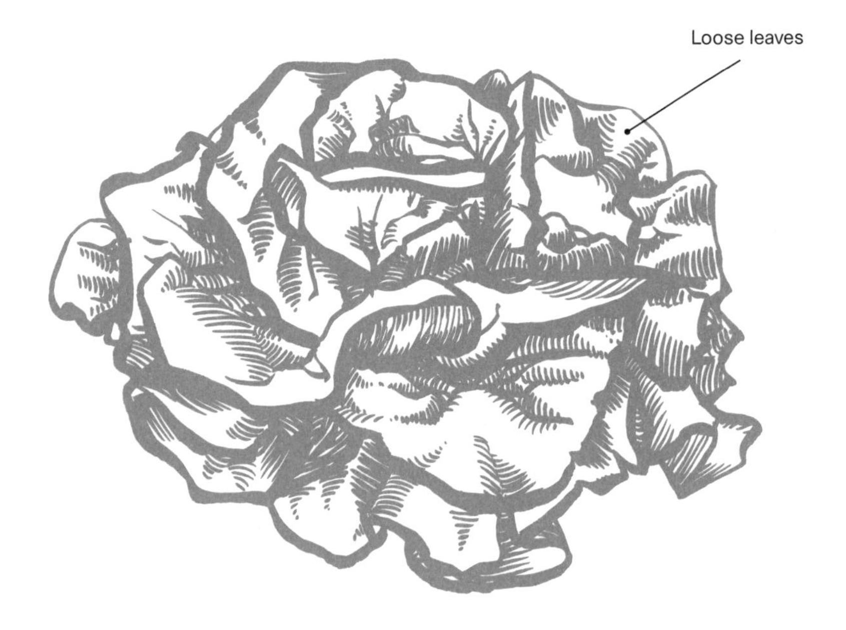

know, as the sweet potato is mainly propagated vegetatively via the production of slips, or young shoots that mainly appear on the proximal or top end, where crown buds exist. These slips are cut off, and as they have small nodes, can form adventitious roots and start new plants.

Ipomoea batatas has several tricks up its sleeve to help it thrive. In addition to smothering competitors with its vining, large-leaved stems, it is an allelopathic plant, which means that it exudes chemicals through its roots into the soil around the plant. These chemicals are taken in by any plants in the nearby area that would compete with the sweet potato, and are affected negatively, often being killed. This allelopathy can be useful for gardeners and farmers, as the crop can be used to control weedy areas, although the allelochemicals do not affect all weeds.

As well as deterring other plants, the sweet potato also defends itself from herbivores through laticiferous (latex-bearing) cells that run throughout the plant, including the tuber. The latex is toxic to herbivores but also to humans, which is why sweet potatoes must be cooked through to destroy the latex. Cooking sweet potatoes has an added benefit, as when they are heated, an enzyme specific to *Ipomoea batatas* breaks down the stored starch into maltose, making the flesh sweet.

Lactuca sativa

Asteraceae

Lettuce

Lactuca sativa is more commonly known as lettuce, a leafy crop that is eaten around the world, most commonly raw, in salads, but also wilted in soups and stir fries. Although it originated in warm climates, it most likely thrived in hilly, cooler areas and it is grown as a cold-weather crop. It will germinate at low temperature, its ideal growing temperature being 13–18°C (55–64°F). If the temperature rises above this for long periods, lettuce will bolt, meaning it runs to seed. Lettuce is naturally an annual or biennial depending on circumstances. When it is ready to seed, the short central stem will elongate, from around 30 cm (12 in) to heights over 1 m (3 ft), where it produces masses of yellow flowers.

Lettuce is a laticiferous crop, producing latex as a deterrent to herbivores. This latex remains at fairly low levels until the plant begins to produce seed, at which point the latex proliferates, rendering the leaves extremely bitter, although they can still be eaten if cooked.

Lettuce is categorised by the way the leaves grow and their texture:

- Heading lettuce, var. *capitata.* These types have short leaves that grow into a compact head that is harvested whole. Ones with soft textured leaves are known as butterhead, and those with more crunchy leaves are crispheads.
- Cos or Romaine lettuce, var. *longifolia*. With long, upright leaves, usually crisp and textured, they have a prominent wide midrib.
- Leaf lettuce, var. *crispa*. This lettuce does not form heads, and instead the loose leaves are picked off individually from the bottom, meaning they can grow for a long period. The leaves themselves tend to be lobed or curly.

Although mainly grown for their green or red leaves, one type is grown for its swollen stem. Known as celtuce, *Lactuca sativa* var. *asparagine* has a thick stem, which has a nutty flavour and is also called asparagus lettuce. The leaves are very coarse and need wilting before eating.

Malus x *domestica*

Rosaceae

Apple

Malux x *domestica* is one of the most cultivated fruits harvested for sale, making the apple an important economic crop. Originally found in Kazakhstan, it was spread worldwide via the trading routes of the Silk Road. The 'x' indicates that the common orchard apple is a hybrid of two or more species, and in this case hybridisation occurred naturally. Some of the parent species include *M. sylvestris*, *M. sieversii* and *M. baccata*.

The fruit of the apple tree is classified as a pome fruit. All pome fruits are members of the *Rosaceae* family and have a core that contains seeds, surrounded by a protective membrane and flesh. Botanically this fruit is also classified as an accessory or false fruit, which seems a bit of an unfair slur. The false aspect comes from the trait that most of the flesh of the apple is derived from the hypanthium, the accessory in question, with only the central core arising from the ovary, which means most of us actually discard the true fruit into the compost bin.

The fruit in this case is created from five carpels, which produce the star shape that is easily observed when cutting across the apple. When handed an apple the top is the end where the peduncle or stalk has been cut away from the tree, and at the opposite end is the eye, which is the remains of the calyx and stamens.

Apples are self-incompatible, resulting in their seeds producing offspring that are genetically different from the parent, a fantastic trait for evolving and thriving in the wild, but not so desirable in an orchard where consistency is key. For this reason, all apples planted are clones, grafted onto rootstocks, which gives control over final height and vigour. As the trees are clones, growers must plant several different cultivars together to achieve pollination and fruit set.

This allogamy, where fertilisation can only be achieved from a different variety, is controlled by the style.

The pollen sets onto the stigma and starts sending out a pollen tube down the style, inducing the style to emit molecules that interact with the pollen, checking its S-alleles (those that control selfing). If the pollen has matching alleles to the style, the growth of the pollen tube is stopped and fertilisation does not take place. When fertilisation does occur, each pollen grain has two sperm leading to double fertilisation. One sperm fertilises the egg in the ovary creating the embryo, with the second sperm meeting the haploid cells in the same ovule, which eventually become the endosperm.

Apple trees are generally diploid, containing two sets of chromosomes, but some are triploid and contain three. Triploid apples such as 'Bramleys Seedling' and 'Blenheim Orange' produce larger fruits and are often much more vigorous, meaning that many ancient apple trees are triploid. But triploid apples have sterile pollen, making them terrible pollinators, a trait that needs considering in orchard management. Rarer still are tetraploid apples, which are useful for pollinating other apple trees as well as producing delicious, large apples.

Crab apples, *M. sylvestris*, are also fantastic apples to have in the orchard as they pollinate abundantly, primarily as they share very few S-alleles with *M.* x *domestica*, thereby avoiding the issue of selfing. But as crab apples are wild and tend to develop traits wanted in domesticated apples, they are being used more in breeding programmes, which will reduce their usefulness as pollinating trees in the future. The fruits of the crab apple are small and very sour, but contain high levels of pectin, a fibre found in plant cell walls that gives structure. Pectin can be used as a thickening and stabilising agent, which is particularly useful when creating preserves such as jam.

Once fruits have begun to form it is normal to observe fruitlets on the ground around the base of the tree. This occurs for six weeks after fruit set, predominantly in June, and in the UK is known as the June Drop. Slightly concerning to the grower, this is a natural process that rids the tree of too much fruit that it cannot support. It

Apple

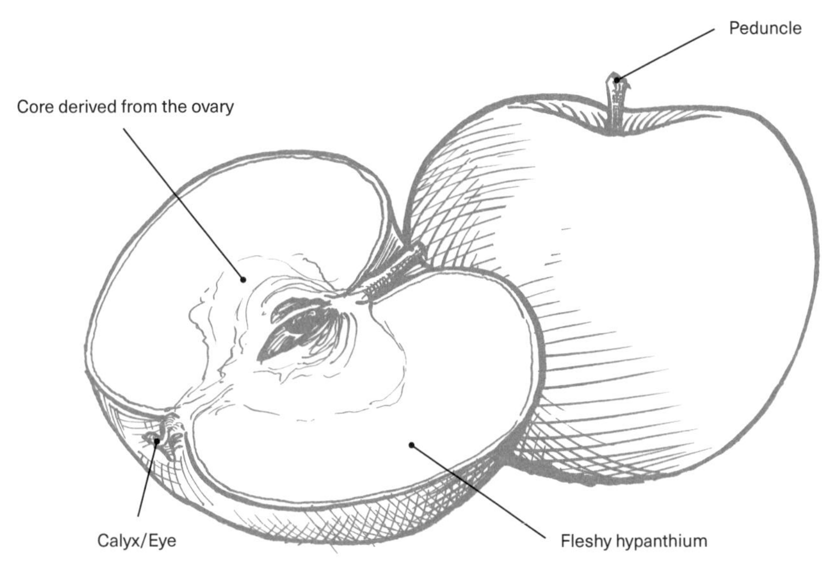

takes around 10 leaves to photosynthesise and produce enough energy to ripen a single fruit, and with human intervention often reducing the number of branches and leaves, the tree needs to ensure it can bring all its fruit to maturation.

The tree will tend towards ridding itself of the fruits with the lowest numbers of seeds, which in the wild leads to more offspring. The seeds within the fruit draw metabolites and hormones to them, and those with fewer seeds owing to poor fertilisation will gain less of these substances. The metabolites, molecules needed for growth, flow over a layer of cells at the base of the fruit stalk called the abscission layer, which is an area of weakness. If there are not enough hormones flowing over this area, it will divide, becoming still weaker and causing the fruit to drop. After six weeks, this layer loses its meristematic abilities, where it can actively divide, and the fruit remains on the tree, unless thinned by human or animal intervention.

Animals play a large part in the dispersal of apple seeds in the wild, particularly bears. Bears are attracted to the colour red, and this colouring also makes the fruits of the apple tree more obvious. Apples are rarely entirely red, with their base colour being green, but are streaked with red. Some modern types have been bred to be entirely red as the consumer relates this to sweetness, a trait humans desire in fruit. The red colour is created by anthocyanins in the apple skin, whose production is triggered by ultraviolet light as one of their roles is protect the tissues from the sun. This is why we often get one rosy patch on the apple where it has faced away from the tree and into the light. Anthocyanin production is dependent on carbohydrates in photosynthesis and glucose metabolism, which is probably why we associate the red colour with sweetness.

The skin of the apple can also be affected by russetting, a brown layer of cells. This occurs naturally in cultivars such as 'Egremont Russet' and produces a sweet and nutty taste. As well as occurring naturally it can be a sign that the apple has undergone stress in the first few weeks of its life, such as cold temperatures or frost, which have affected the epidermal cells. To heal the damage the apple forms a scab, a corky layer of suberin, which is a fatty substance that waterproofs cells and stops decay.

During the summer, while the fruits are growing, the tree is producing the buds that will become next year's fruits. These buds, along with the rest of the tree, need to become dormant over winter. Once the tree drops its leaves, it cannot photosynthesise again until the spring, meaning it cannot waste its stored energy, hence the dormant period. To enter dormancy and exit it, the tree needs a set number of winter chill hours, which for most apples is 500–1,000 hours consistently below 7.5°C (45°F), although there are some low-chill types that only need 300, such as 'Fuji' and 'Pink Lady'. Without sufficient chill hours, the tree may have reduced fruit set and poor-quality fruits the following year.

Morus nigra

Moraceae

Black mulberry

These small, black fruits hang off the deciduous tree known as the black mulberry, or *Morus nigra*. This is a large tree, reaching 10 m (32 ft), which readily branches, forming a dense canopy. Mulberries are known to be home to silkworms, which eat their leaves, but sadly, they are much fonder of the white mulberry, *Morus alba*, than *M. nigra.*

The fruits of the black mulberry grow in the axils of current season growth, or the spurs that come off older wood. They start life as monoecious, pendulous catkins. Catkins are elongated clusters of flowers, or inflorescences, that generally lack petals, but do have scaly bracts and are wind-pollinated. The catkins on the mulberry are short, green and non-showy. They are wind-pollinated, but can produce fruits without being pollinated at all.

The black mulberry fruit is a multiple fruit, or a syncarp, as it is made up of more than one ovary, and these smaller units can be seen upon closer inspection. As a true fruit is made from the ovary, the actual fruit part of the mulberry is an achene, which is the hard, dry seed inside each small unit of the multiple fruit. The fleshy part of a mulberry is made from the four sepals that enclosed the ovary when it was at the catkin stage, so, like the strawberry, it is an accessory fruit.

An achene enclosed by flesh can be categorised as a drupe, which is a central hard stone or seed covered by skin, the exocarp and flesh, or mesocarp. When lots of drupes form to make a multiple fruit, they are called drupelets. So, confusingly, botanically speaking mulberry is not a berry at all.

Musa spp.

Musaceae

Banana, Plantain

The fruit of the banana is enjoyed around the globe, its creamy yellow flesh being a staple of the fruit bowl. Most bananas cultivated for harvest are hybrids, and do not belong to a specific species, and in this entry we mainly refer simply to *Musa* spp.

The banana plant strongly resembles a tree, as it is up to 10 m (32 ft) and has a single, central trunk. But it is in fact a herbaceous plant, with the

Mulberry

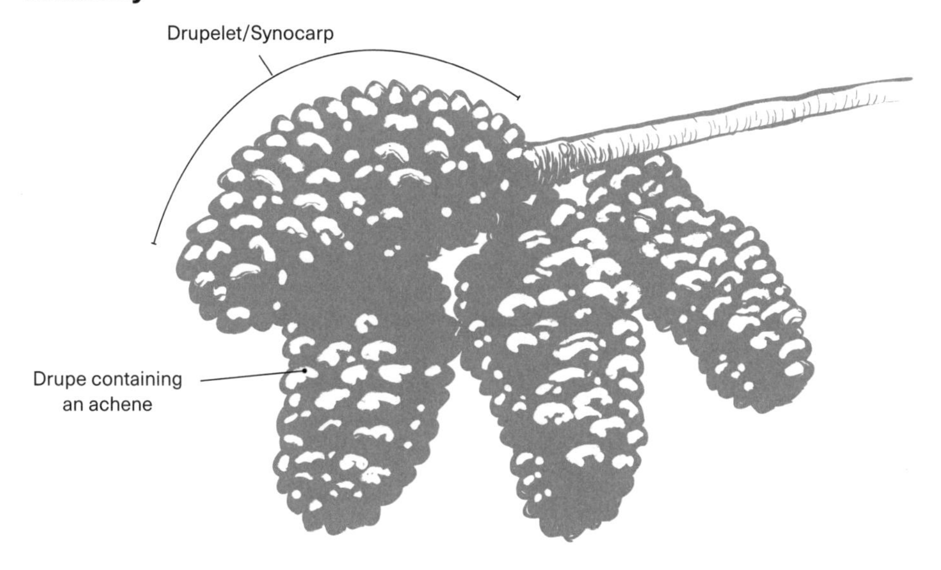

Banana

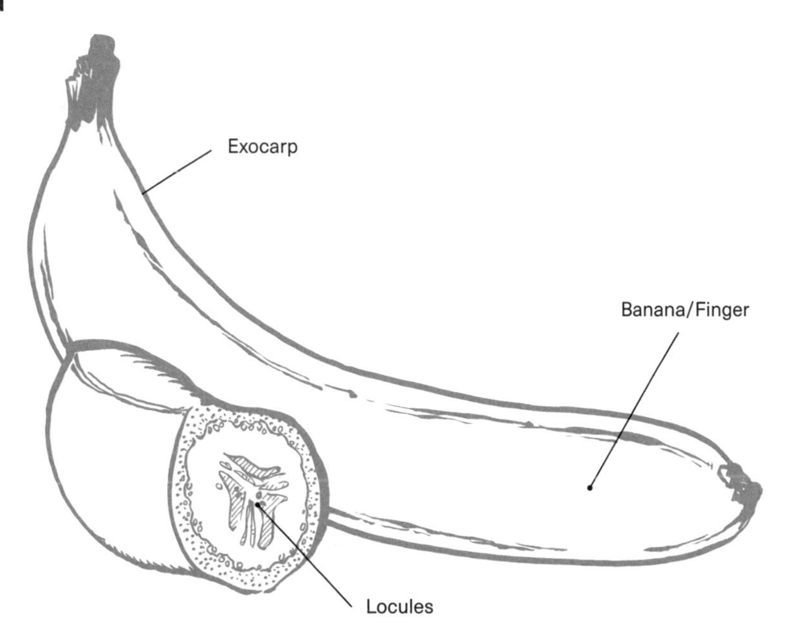

The botany of training fruit trees

Most fruit trees can be trained and pruned into smaller styles, lending themselves to urban growing spaces.

Left to their own devices, most fruit trees will grow to large sizes, even if on dwarfing rootstock, but many growers only have a relatively small space to produce fruit. Most fruit trees can be trained and pruned into smaller styles, lending themselves to urban growing spaces. Using restrictive pruning, these fruit trees are generally grown in one plane, often against a wall or fence, and although yielding smaller amounts of fruit, are fantastically productive for their size.

The restricted forms have several different names depending on their end result, but are based on three main training systems: the cordon, which is a single, central stem; the espalier, where the central stem is allowed to produce lateral branches for fruiting; and the fan system, which has several main arms with laterals growing off them.

To get healthy trained fruit, ensure they are on the most dwarfing rootstock possible so that the tree is not constantly fighting against the restraint of dwarf pruning. Restricted fruit is always pruned in summer to control growth. Pruning restricted fruit in winter leads to a mass of new stem production in the spring, which is undesirable. When a tree goes dormant for the winter, it sends its reserves to the roots, just enough for the volume of the aerial parts. If, during the winter, parts of the tree are removed, there are more reserves than needed, so in spring this bursts through buds, creating new stems. By not pruning in the winter there is the right volume of reserves for the tree, creating steady growth in spring when the sap starts to rise.

Most restricted fruit aims to produce a system of horizontal branches to encourage fruiting.

Vertical branches tend to be extremely vigorous but remain vegetative. This is due in the main to the hormone auxin, and the effect of apical dominance. Lowering branches to become more horizontal reduces the effect of apical dominance, sending more auxin to the lateral buds where fruit will form. Bending branches also produces more of the hormone ethylene, which is involved in bud formation and the creation of some flowering tissues. A branch of 30° above vertical is thought to be the perfect angle to produce the most fruit.

Two other ways to manipulate a trained fruit tree are called nicking and notching, and are used to control which buds break into stems.

Two other ways to manipulate a trained fruit tree are called nicking and notching, and are used to control which buds break into stems. Notching encourages a particular bud to grow, meaning if there is a large gap in growth on the restricted fruit form the grower can fill it. A shallow notch is carved just above the target bud, generally on the main stem, disrupting the flow of auxin as it heads down the stem, suppressing the growth of the lateral buds. In this way, the bud is able to grow out strongly. Nicking is the reverse, where a shallow cut is made just below an undesirable bud to stop the upward rise of sap to that particular area, preventing it growing out into a stem.

central trunk made up of overlapping, clasping leaf bases, which form a solid pseudostem. The true stem is underground and is a storage organ, or rhizome: a swollen, horizontally creeping stem. The banana is a tropical plant and will not grow in all kitchen gardens, as it requires humidity, but a glasshouse may provide suitable conditions

The flowers on the banana plant appear on a central flower stem, with the males at the very end, and the females further down. The flowers are protected by a large red bract known as a spathe. The male flowers are generally sterile, and simply die off, meaning most banana fruits are produced through parthenocarpy, with no pollination. For humans, this means that the bananas are seedless, which is advantageous, because when pollination does occur, the seeds are extremely hard.

The downside of this is that, as bananas are herbaceous and generally die back once fruited, with no seeds, the only way to continue the banana line is through vegetative propagation using parts of the rhizome. This produces clones of the parent plant, which give a perfect banana, but little resistance if a disease starts to attack the crop.

The fruit itself hangs off the stem in clusters, which are made up of bunches, known as hands, which in turn are made up of individual bananas, known as fingers. A bunch can be made of up to five hands, each with five to 20 bananas, so each plant produces a lot of fruit. The skins can be green, red or yellow, although generally start life as green and ripen. The peel is the exocarp, which is thickened with pulp created by the receptacle. When removing the peel, the stringy parts that come away were formed by the xylem and phloem, which leaves the fleshy inner. The banana is made of three chambers or locules, which can sometimes be seen if the flesh is squashed slightly. Inside this are a series of three dots, which are the aborted ovules where the seeds would have been.

Botanically, the fruit is a pepo, a subcategory of a berry. A berry is any simple fleshy fruit formed from a single ovary of a single flower, and a pepo is a berry with a hard rind. The banana is renowned for its slightly curved shape, and this is due to negative geotropism. The flower spike bends down when weighted with fruit, but the fruits themselves experience negative geotropism and grow away from the ground, searching for sunlight. This upward motion is opposed by gravity, which all leads to a slight bend in the fruit's shape.

As well as the sweet dessert bananas that are eaten raw and have a high sugar content, there are more starchy types of banana known as plantain. These are commonly cooked and eaten as part of a savoury dish.

Nasturtium officinale

Brassicaceae

Watercress

Nasturtium officinale rather confusingly shares its Latin name with the common name of a completely different plant. The flowers that grow in many kitchen gardens are commonly called nasturtiums, but their Latin name is *Tropaeolum* spp., and although they are edible and have a similar taste to that of *Nasturtium officinale* they belong in a different family altogether. The common name for *N. officinale* is watercress, a leafy crop with the sharp mustard flavour that is synonymous with the brassica family.

As the common name suggests, watercress is a semi-aquatic plant, which thrives in either very moist soils or clear, slow-running water, although it can be grown in a pot of water in the kitchen garden. It is a perennial but is usually grown as an annual and is cultivated for its leaves and young shoots. The leaves are dark green and pinnate in shape. The dark green colour

Watercress

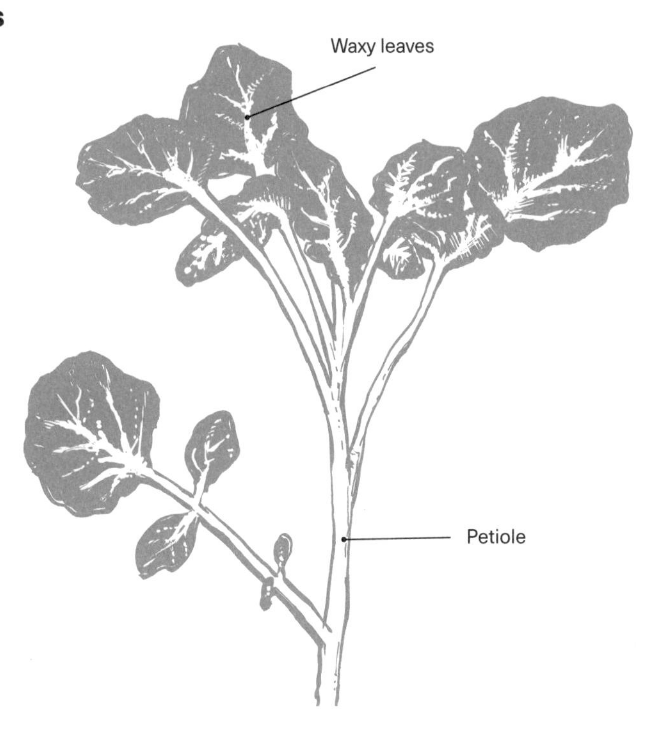

indicates high presence of chlorophyll, a green pigment that is used as part of photosynthesis. High chlorophyll in turn means this leaf is full of iron, as plants use iron in the formation of chlorophyll, hence we are often told to eat our greens for a healthy diet.

Watercress is native to Britain and parts of Europe and grows well at cooler temperatures. It is triggered to flower through photoperiodism, when day length is more than 12 hours. At this point in the year it will flower, so often finds itself constantly attempting to go to seed during the summer. If flower heads do appear, they can be pinched out, which promotes a bushier plant, as apical dominance is removed. In cultivation, *N. officinale* is known as green watercress, which stays green throughout its life cycle, but is not very frost-hardy. Alongside this is often grown *N. microphyllum*, known as brown watercress, which is more hardy, but the leaves turn a purple–brown in the autumn.

An aquatic plant, watercress has a few adaptations to live in wet conditions. It has a waxy epidermis on its leaves, which although in many plants stops water loss, in this case stops waterlogging of the cells. Watercress also has hollow stems, which allows the top, green parts of the plant to float. Within the roots, the parenchyma cells will modify to become aerenchyma cells, especially if conditions become very hypoxic or oxygen-depleted (there is a good supply of oxygen in running water or soil in normal conditions).

Aerenchyma cells are spongy with large cavities for gas exchange. As the roots are unable to bring in oxygen as they normally would, the aerenchyma allows gas to flow more freely from the leaves down to the roots, and also allows roots to get rid of carbon dioxide. These cells also take the place of cortex cells, which reduces the need for energy production. The watercress will put out adventitious roots at its leaf axils, which also increases the number of aerenchyma cells in the plant.

Olea europaea

Oleaceae

Olive

The olive tree, *Olea europaea*, is a xerophytic plant, meaning it is adapted to live in places with very little available water. The striking, narrow, silver leaves of this evergreen tree are one obvious modification. Having a narrow leaf means less surface area for transpiration, conserving water within the plant. The silver colour reflects the heat of sun, which means the plant spends less energy trying to evaporate water and cool the leaves.

Olives thrive in Mediterranean climates with low rainfall, warm summers and cool, but not frozen winters. The tree itself can withstand cold temperatures down to −10°C (14°F), but the fruit cannot, which is why, in cooler countries, olives are often stored in glasshouses overwinter.

The tree itself can grow to 15 m (50 ft), but in most situations is kept to 5 m (16 ft) to allow easy harvesting. The flowers grow in panicles or racemes, of perfect flowers. The trees are generally self-sterile so need to be cross-pollinated for successful pollination and fruit set. Olives are anemophilous, relying on the wind to transport pollen.

Once pollinated, the flowers produce a single-seeded, indehiscent drupe. The part that humans are interested in is the mesocarp or flesh, which surrounds the seed. The olive stores food as oil, unlike many other fruits, which store it as carbohydrates. The oil is stored in the cytoplasm of the mesocarp cells, and can be up to 75% oil. Not all olives are quite so high in oil, as some are bred for oil production and others as table olives, which have less oil.

Olive

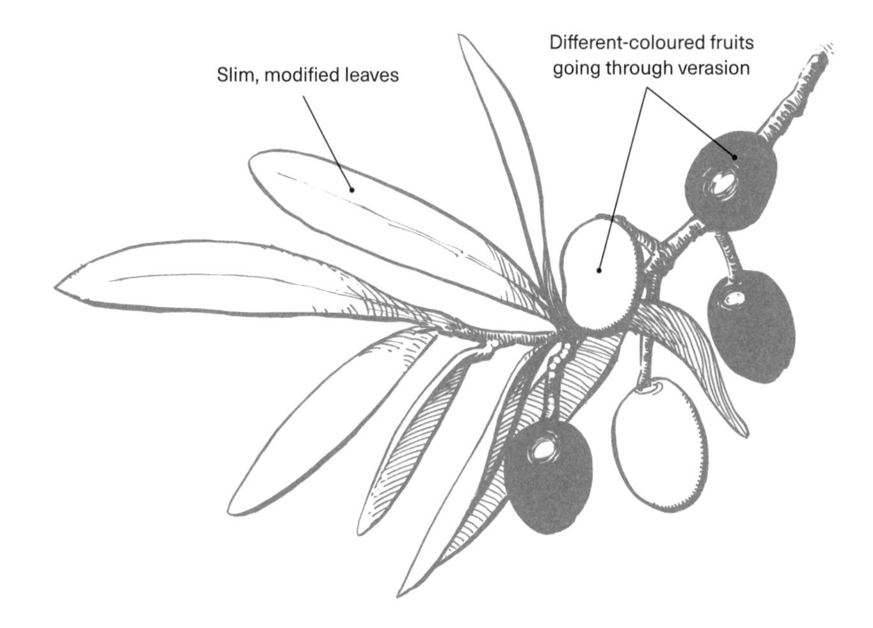

Olives start life as a green colour, which ripens through blue, purple, red and black, a process called verasion or colour change. Green olives contain a bitter compound called oleuropein, which is removed after picking by neutralising with sodium hydroxide and then pickling in brine.

The seed of the olive is surrounded by the endocarp, which is too hard for human teeth to break, and is generally discarded. The endocarp is hardened up to protect the seed via a process called sclerification. This is the secondary thickening of the cell walls by lignin, which eventually floods all the cells, killing the living material and forming a thick protective cell. In cultivation this hard wall needs to be broken down via scarification, the physical breaking of the wall, or with acid to allow the seed to germinate. In the wild this occurs naturally, in the guts of the birds that would have fed on this crop.

Passiflora edulis

Passifloraceae

Passionfruit, Granadilla

Passiflora edulis is a tropical, woody vine that can reach heights of 15 m (50 ft). It scrambles over other plants and hoists itself along using tendrils that grow from its leaf axils. An evergreen plant, it is not hardy and needs a long, warm season to fruit, so is not always suitable for colder climates, where it is more likely to be grown for its flowers.

The flowers are beautiful, in purple and white with a prominent centre and frilly edge. The column in the centre has a triple-tipped style at its pinnacle, and at its base is surrounded by a dark ring known as the operculum. The operculum sits over the nectaries and prevents nectar-robbing. Surrounding this are the frills, which are in fact corona filaments and have

Passionfruit

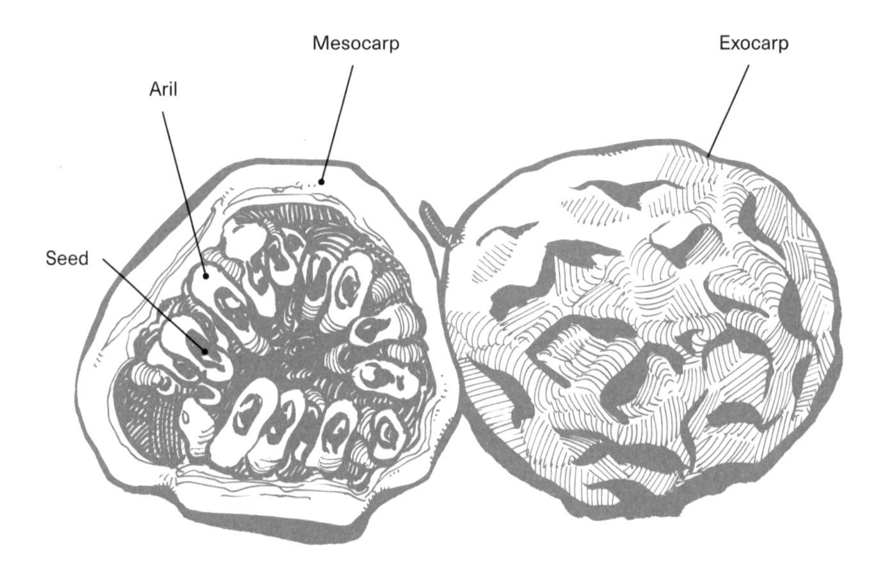

varying concentric rings of colour, generally purple and white, which point pollinators towards the nectaries.

Once pollination occurs, the fruit of *Passiflora edulis* is formed, commonly known as a passionfruit or granadilla. Straight species *P. edulis* forms purple passionfruits, which have a hard, wrinkled outside rind. There is also the yellow passionfruit with a softer, yellow rind, which is *P. edulis* var. *flavicarpa*. Both have a hard outside rind, with a layer of white pith underneath the rind and are filled with flesh, which makes the passionfruit a pepo berry.

Within the rind are up to 250 hard, black seeds, which are each fully surrounded by arils. An aril is a fleshy appendage that often surrounds seeds and contains sweet flesh to attract frugivores, which will eat the aril and seed and disperse the plant further afield. In the case of the passionfruit, the aril is a membranous sac that is filled with juice, which as well as attracting animals for distribution purposes, is also thought to give the seeds some physical protection when ripening.

Pastinaca sativa

Apiaceae

Parsnip

The parsnip is a stalwart of the kitchen garden, sown in spring and growing slowly throughout the summer and autumn, finally to be harvested as a sweet, winter root crop. *Pastinaca sativa* grows wild in southern and central Europe, where it produces a small, woody root, whereas its cultivated cousins have a large, fleshy root that can be up to 10 cm (4 in) in diameter at its

Parsnip

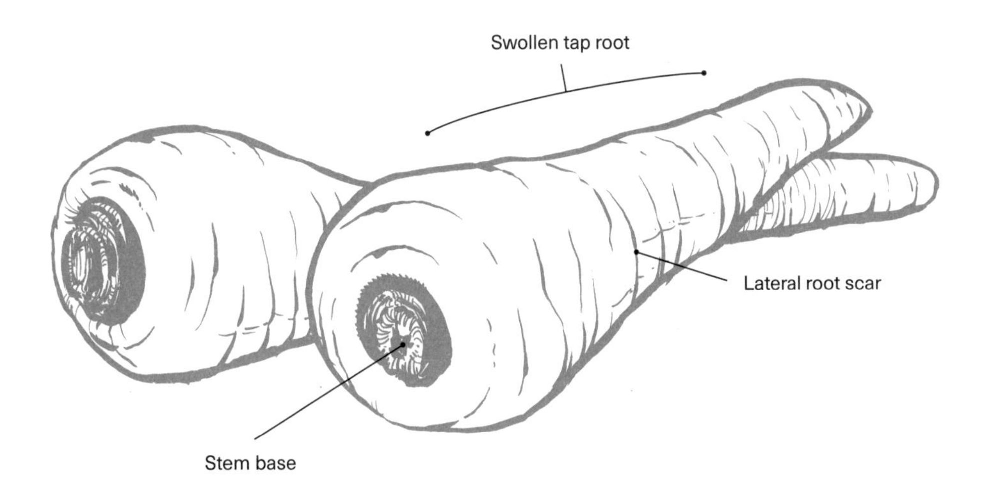

shoulders, and up to 50 cm (20 in) long, although much of that length is the slim end of the tap root, that usually snaps off during harvesting.

The parsnip is a biennial, which produces seeds in its second year of growth, but as it is generally harvested in the winter before this occurs, it is often referred to as an annual crop, especially as the core becomes very woody when it is about to flower. Above ground, parsnips have large, pinnate leaves on hollow petioles. The stem is not elongated, but instead is modified to become a flattened disc, which lies at the top of the root crop and bears buds that produce the leaves. Once the herbaceous plant does produce seed, it dies, making it a monocarpic plant.

Parsnips are true root crops – it is their swollen tap root that is the main harvest, although the top part of the parsnip is the swollen hypocotyl and lies just below the crown. The parsnip will often have small, lateral roots growing sideways from horizontal scars, and these gather water from the soil. As with its cousin, the carrot, a cross-section of the parsnip shows a central xylem system and an outer phloem ring, just under the skin. Between them lie the cortex cells, made of storage parenchyma and containing starches and sugars.

Generally considered a winter crop, parsnips are said to become sweeter after a frost. When hardy plants experience frozen temperatures, they will convert some of their stored starches to sugars. These sugars act as an anti-freeze in the cells by raising the concentration and lowering the temperature at which the water freezes. If the cells do freeze, they can be so fragile in the frozen state that they break and die. Many plants create this sugary anti-freeze and taste sweeter after a frost.

The botany of forcing crops

Crops in the kitchen garden grow, mature and ripen based on cues from the environment around them, predominantly light levels and temperature. Plants can be tricked into producing a crop outside of their normal growing season, a technique known as forcing.

As well as forcing plants by artificially creating the right conditions, crops can also be manipulated to be sweeter and more tender by taking away light.

In some cases, it is as simple as growing indoors and giving the plants a little extra heat, encouraging them to ripen a few weeks earlier than their outdoor counterparts, as is the case with strawberries. Supplying artificial lighting along with the right temperatures can also trigger fruit or flower production out of season.

Some plants measure the amount of dark in a day, a trait called photoperiodism. In simplified terms, they use a pigment in their leaves called phytochrome to measure the amount of red light, and can determine how long the night period is from this. Although the plant measures the night length, growers talk in terms of day lengths. At specific day lengths, called the critical day length, some plants are triggered to start certain processes, such as flowering and fruiting. As the same day lengths occur twice a year, this trigger is also often dependent on a temperature cue as well, such as the cold of the winter or warmth of the summer. Growers can therefore persuade plants to fruit outside their season if they can give the right temperature and the right light levels, but this is expensive and uncommon in the kitchen garden.

As well as forcing plants by artificially creating the right conditions, crops can also be manipulated to be sweeter and more tender by taking away light. This is also called forcing, and

As forced plants do not photosynthesise, their colour tends to be much paler.

sometimes blanching. Crops such as rhubarb, seakale and chicory are traditionally forced to produce out-of-season crops, which are sweeter and more tender than when growing in their natural season. In this type of forcing the plants are given heat but all light is removed.

Growing crops in the dark has several effects on the plant tissue. It cannot photosynthesise, so must draw on reserves for energy to grow. When drawing on reserves of metabolic activity, the plant must first convert the stored starch into sugars, which creates a sweeter crop. When light is removed from a crop, it becomes etiolated, or stretched out as it searches for the light. Etiolation produces cell walls that are much weaker – a bad thing for the plant, but making a much more tender harvest. Etiolation offers faster growth, especially if there is warmth given to the crop, as is generally the case with rhubarb and chicons.

Because forced plants do not photosynthesise, their colour tends to be much paler. In some crops, such as leeks and celery, parts of the plant have light removed to give paler, sweeter, more tender areas. Leeks are often mounded around the base, to give a long white shank, and celery is grown closely together to encourage the fight for light and taller, more tender petioles.

Phaseolus spp.

Fabaceae

Beans

The genus *Phaseolus* includes some of the most widely cultivated beans. The majority of the species originate in Central and South America and are generally tender plants that cannot withstand severe frost. They are herbaceous plants, and quite often have adapted to be climbing vines, a fantastic evolutionary trait that allows them to scramble up and over other plants to gain as much sunlight as possible.

They are commonly known are legumes, due to the fruit they produce, which is botanically called a legume. A legume fruit is a simple, dry fruit, that dehisces, or splits open along a seam, and in the case of legumes, along two seams. There are other genera classified as legumes, such as *Pisum* and *Vigna.* Most legumes also have a symbiotic relationship with nitrogen-fixing bacteria that live within their roots.

There are types of nitrogen in the soil that are unavailable to plants, but the bacteria can convert them into available forms, in exchange for energy from the plant. Nitrogen is important to plants as it forms proteins and amino acids. An abundance of nitrogen also means that the food reserves of legumes comprise protein rather than starch, which is why beans are such an important crop. Beans are either harvested when immature, when the pod and seeds are generally consumed, or when they are mature and dried and only the seeds are eaten.

The body of the pod of *Phaseolus* plants is formed from the ovary and has two distinct ends. At the top, where the pod joins to the plant, which, after harvest, usually consists of the stalk or pedicel, two green bracts come away from the pod like wings; there are the remains of the petals that have dried away to form a hard ring. At the opposite end, the pod almost forms a point, known as a beak, and is the elongated stigma and style.

If the pods are harvested when immature to be eaten whole, they will often contain tiny seeds that are not yet dried. These are attached to the pod via the funiculus, which acts as a placenta, feeding the seed food and water from the plant. If harvested when mature and dried, the funiculus has disappeared, but leaves a scar in the concave side of the bean, known as the hilum.

The hilum of beans is often a different colour from the rest of the surface and is quite noticeable. Alongside the hilum, a tiny hole can often be seen, known as the micropyle. The micropyle is formed when the seed is still an ovule, and is the opening where the pollen tube enters. Once dried, the hole still exists, and is where the first roots or radicle emerge when the seed germinates. The radicle can be seen on a bean seed before germination, as it forms a ridge just to the side of the hilum. On the other side of the hilum to the radicle is another ridge, known as the raphe, which is simply the shape formed where the seed sat against the funiculus. The hilum has a tiny growth to the it, known as the strophiole, which controls water in and out of the seed.

The entire seed sits within a seed coat, called the testa, which will often come away from the bean once cooked. The majority of the bean seed is made of the cotyledons, which will emerge upon germination as the first leaves. Bean seeds are what is known as non-endospermic. In endospermic seeds, the plant stores food for the embryo in the endosperm. In non-endospermic seeds, the endosperm is absorbed by the cotyledons and makes them large and fleshy, ready to create the new plant.

One slightly undesirable trait in beans is their string, which forms as they mature, down the seams of the pod. The string is a protective lignin cap that runs down the phloem vessels, formed by sclerenchyma cells, which are dead cells that

Green bean

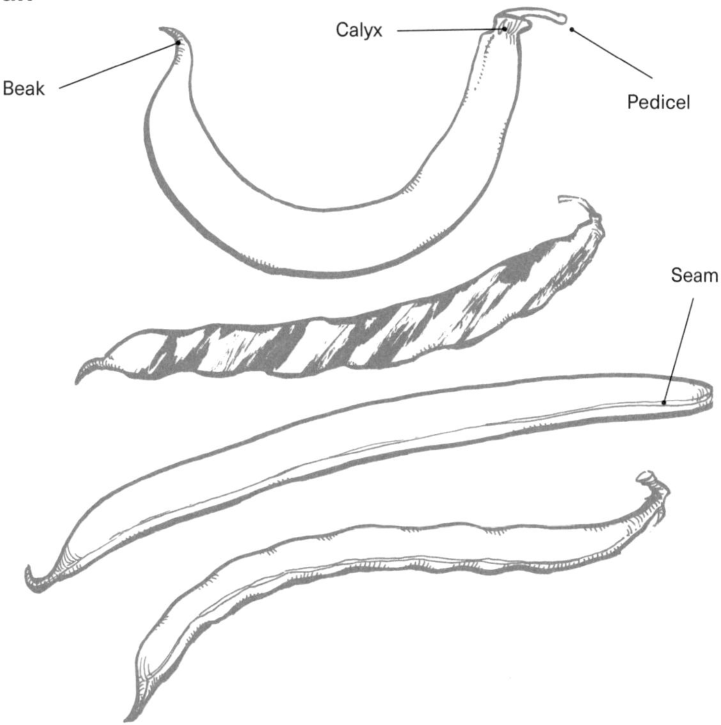

act as support. These strings are also involved in the dehiscence process, helping the dried pod twist and split open, dispersing its seed.

There are three main species within the *Phaseolus* genus that are cultivated in the kitchen garden: *P. vulgaris*, the common bean; *P. coccineous,* the runner bean; and *P. lunatus*, the lima or butter bean. The lima bean is less commonly grown in cooler climates, as needs a little more heat, and is not covered here.

Phaseolus vulgaris

Common bean, Green bean, Haricot bean, Kidney bean, French bean, Borlotti bean

Phaseolus vulgaris goes by many different names, mainly due to massive variety in the seeds, which when dried have names such as haricot, kidney and borlotti bean and vary in size and colour, from dark black to pure white, but all usually bearing a white hilum. But common beans are also widely harvested when green and the whole pod is eaten, at which stage they are commonly called green or French beans. The pods are generally green, but can also be yellow or purple, and may be speckled, as is the case with borlotti beans, which have red blotches over the pod and are very attractive.

The common bean is an annual, and grows well at temperatures around 25°C (77°F), but does not tolerate frosts – in fact, consistent temperatures below 8°C (46°F) will stop flower production. The common bean is either a bush

Bean seed

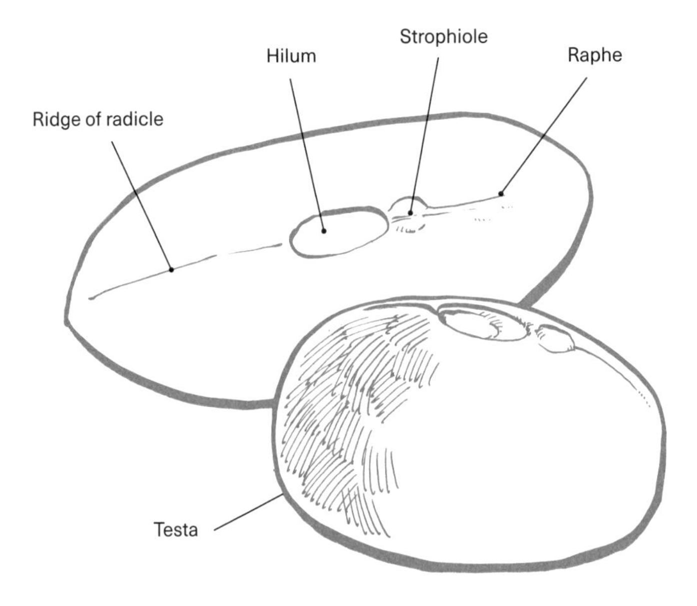

or trailing type. Bush beans are also known as dwarf beans and only grow to around 60 cm (24 in). They are determinate, so only grow to this height and only produce a certain number of flowers, but they need little support and are useful for smaller spaces. Trailing beans are also known as pole beans, and are long vines growing to 3 m (10 ft) and continuously produce flowers in their leaf axils.

These types were domesticated long ago alongside taller plants, most famously in a system called the three sisters, in which maize, beans and squash were grown in a small space and then harvested once dried to store for winter. In this system, the beans grow up the tall maize, and the squash expands around the ground, adding shade to the roots and suppressing weeds. From systems like this, the training beans over time learnt to grow upwards, whereas this wasn't advantageous to bush beans, which were not domesticated in these systems.

Phaseolus vulgaris is self-pollinated, which has the advantage that the seeds saved will come true to type if planted again the next year. The system of self-pollination is an introrse stigma, which means it is twisted to face inwards and directly receives pollen from the anthers.

Phaseolus coccineus

Runner bean, Scarlet bean

The runner bean is a large vine, reaching heights of 4 m (13 ft), and the pods and beans produced are also much bigger than those of the common bean types. Runner bean pods, like the common bean, are harvested either immaturely or when dried for the seeds. Traditionally, this plant was often grown as an ornamental for its striking flowers, which were either vivid red or pure

white, although new cultivars exist with pink and salmon flowers.

Although it is a warm-season plant, if night-time temperatures persist above 15°C (59°F), with high humidity, the plant will abort these flowers, as in the wild these do not represent the best conditions for flower set to occur. The flowers are most likely to be cross-pollinated, as the stigma in *P. coccineus* is extrose, which means it is outward-facing and pollinating insects such as bees can introduce pollen from elsewhere.

Runner beans are perennial plants in the wild, and as such have an extremely large, fleshy tap root, which stores reserves for the dormant period. In some cultures, the root is also eaten. Runner beans are more traditionally grown as annuals in the kitchen garden, especially in colder, wetter climates where the roots will rot off, but where treated as a perennial, the pod production in the second year is often earlier, but not so vigorous.

As is the case with all beans, constant picking produces more beans. The aim of any plant is to produce the next generation, and when the seed is removed, the plant needs to produce more flowers and pods to complete its mission.

Physalis philadelphica

Solanaceae

Tomatillo

A native of Mexico, the tomatillo is a crop grown for its edible fruits, which are predominantly used in savoury dishes such as salsa. It is in the same family as the tomato and has similar flowers and taste. The tomatillo is a herbaceous, annual plant, which grows to around 1 m (3 ft), but tends to flop over and become prostrate. It has slightly hairy leaves, which help reduce water loss via transpiration as the hairs create a small, humid microclimate, which evens out water concentrations – a big bonus when growing in warm climates.

The flowers of the tomatillo are yellow with purple blotches in the centre and produce fruits that are botanically berries and contain hundreds of seeds. The fruits can be green, yellow and orange, and have a sharp, slightly citrusy flavour. The tomatillo itself is not hardy at all, and cannot live through frosts, growing best at higher temperatures.

A characteristic sight on a mature tomatillo plant is the papery lanterns or husks that hang

Tomatillo

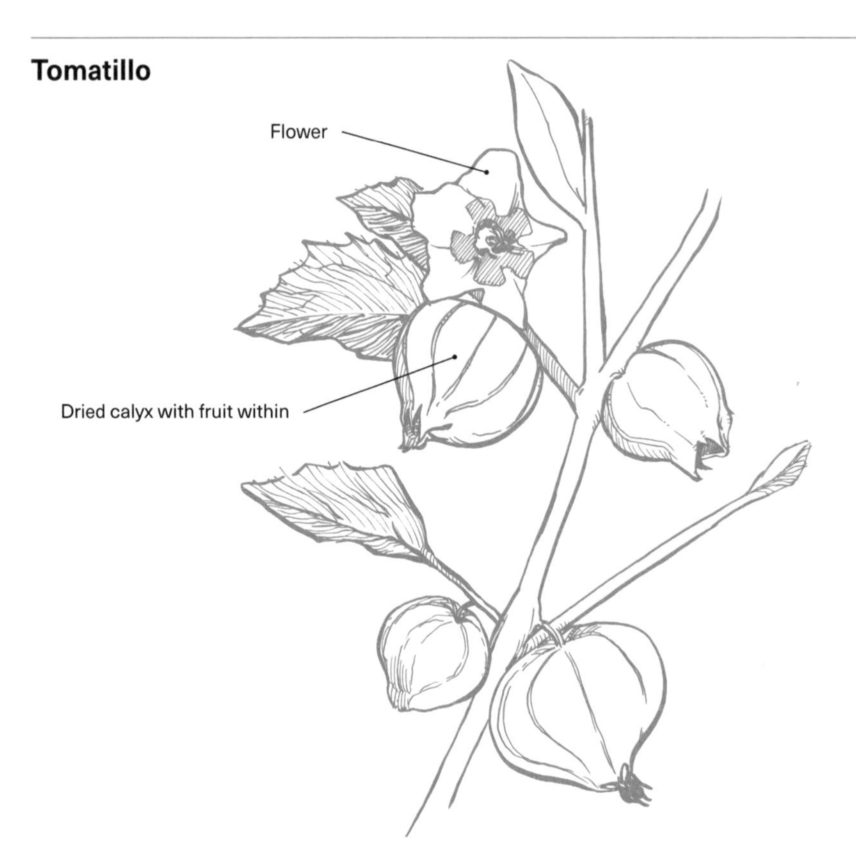

from its stems. These husks hide the fruits from sight until they are mature and break through the paper walls. The husks are formed from the calyx of the flower. The calyx is the collective name for the lower whorl of the flower, comprising sepals, which are generally green and protect the flower when in bud.

Once the tomatillo has stopped flowering, the calyx continues to grow, becoming what is known as an accrescent calyx, and eventually completely envelops the fruit. This husk protects the fruit from predators such as birds and insects and keeps away diseases and the effects of extreme weather. It also acts as a source of carbohydrate during the first twenty days of fruit development.

Physalis peruviana

Cape gooseberry

P. peruviana goes by many names, including Cape gooseberry, ground cherry and goldenberry. Unlike the tomatillo, the fruits of the Cape gooseberry are slightly sweet, although still tart, and are more typically harvested for the fruit bowl and used in desserts. The plant is an evergreen shrub in warm climates, but is more typically grown as an annual herbaceous plant, as it cannot tolerate cold weather. Growing up to 1.5 m (5 ft),

it is taller than the tomatillo, although again generally spreads around. But its the flowers and fruit are much smaller. The fruit is hidden from sight by the papery calyx husks, but is bright orange, hence the common name of goldenberry.

Pisum sativum

Fabaceae

Pea

Domesticated in south-west Asia and the eastern Mediterranean, *Pisum sativum* is a versatile food plant, with edible flowers, shoots, fresh and dried seeds and in some cases, edible pods. The pea is an annual, cold-climate plant that grows extremely well around the 20°C (68°F) mark, but happily survives through periods of freezing. It has a tendency to die if exposed to prolonged periods at temperatures above 25°C (77°F).

The overall plant can be quite variable, from indeterminate, with flowers along the stems and in the axils, to determinate, with bush habits where flowers are borne on the terminal bud. The flowers themselves are generally white and have a distinctive shape, with five petals, of which two are joined to make the keel, which protects the reproductive organs. The keel is just within the petals known as wings, and the largest petal is called the standard.

Peas are cross-pollinated, a trait that was famously used by the monk Gregor Mendel in the mid-1800s to demonstrate the principles of genetics through dominant and recessive genes.

The pea pod Is a simple fruit, classified as a legume, with the pod itself a swollen ovary and the peas the developed seeds. On the pod, the peduncle and sepals can still usually be very clearly seen, and often, at the opposite end, the style and stigma are still visible. The pod has a protective function, and as well as transporting nutrients to the seeds, it photosynthesises to harvest energy from sunlight.

The pod of the common garden pea has distinct layers: the outer layer contains the chloroplasts where photosynthesis takes place, and the inner is a fibrous parchment layer. As it matures, this fibrous layer dries out and stiffens faster than the outer layer, which causes the pod to twist and split along the seam, dispersing the

Pea

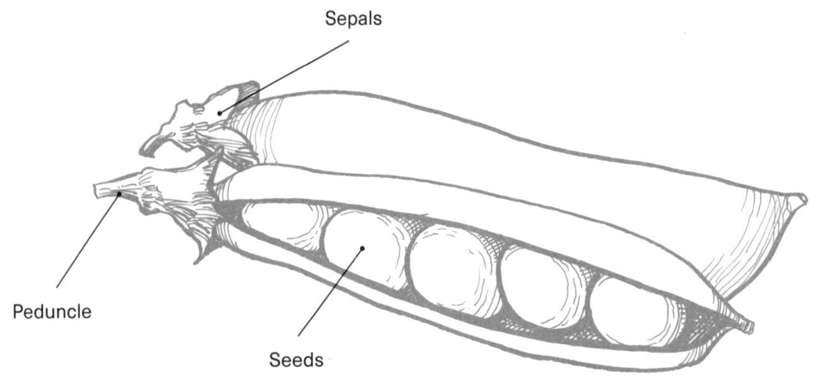

Pea flower

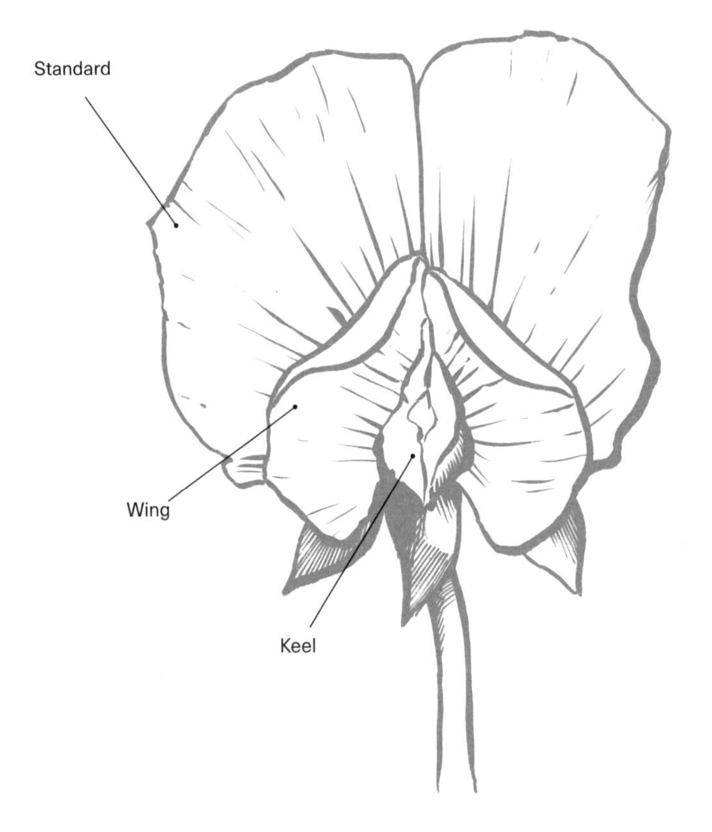

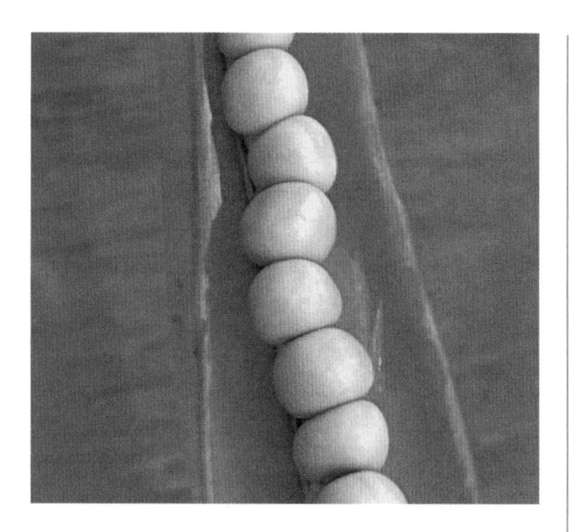

seeds, categorising the fruit as dehiscent. This fibrous layer also makes the raw pod inedible, although it can be cooked and used for stocks.

The pea is the seed of the plant and is often harvested at an immature stage while still soft for eating fresh, or at a later stage when dried and mature for storing. The seeds themselves can be green, yellow and occasionally white. The appearance of a pea is defined as wrinkled or smooth, although genetically is it now thought there are four types, three wrinkled and one smooth.

The wrinkled types taste sweeter owing to a higher sugar content. This high sugar concentration causes more water to flow into the seed, via osmosis, which makes them swell up more. While immature, there are no visible differences, just the taste. Once they start to dry out, the wrinkled peas have a relatively higher amount of water to shed, and because the skin has stretched out so far, it visibly wrinkles when shrinking back on maturity, whereas the smooth peas have less water to lose and therefore the skin is not stretched.

Peas and other pulses, mature seeds, are renowned for being a good source of protein. As the plant creates symbiotic relationships in the soil with nitrogen-fixing bacteria, it has excessive nitrogen to create and store amino acids, the building blocks of proteins.

Pisum sativum has leaflets on its stems, but the adaptation of these to become tendrils allows the plant to climb and hold on to supports. The tendrils achieve the ability to twine around garden canes via thigmotropism, a curling in response to touch. Once the tendril senses an object, several processes take place to cause the movement. One process involves the hormone auxin, which among other roles is responsible for cell elongation. Once the plant senses an object, auxin elongates the cells on the opposing side of the tendril, creating a curling motion, so the tendril can hold on and support the plant.

The ability to climb above other plants is advantageous in the wild as it allows more access to sun and therefore more energy produced from photosynthesis. Some pea plants have excessive tendrils and almost no leaflets; these are known as Afila peas, and the tendrils can be harvested and used as edible decoration.

Pisum sativum **var.** ***macrocarpon***

Mangetout, snow peas, sugar snap peas, Chinese pea

Unlike common garden peas with their inedible parchment walls, *P. sativum* var. *macrocarpon* are harvested as an entire pod as they lack a fibrous layer in the pod. They commonly have pink or purple flowers instead of white. Known as mangetout, or 'eat all', they can be subcategorised as snow peas and sugar snap peas. Snow peas, also referred to as the Chinese pea, originated in south-west Asia and are extremely hardy; they are harvested when the pod is flat with very tiny seeds. The sugar snap pea is harvested when the peas are bigger, giving a fuller pod, but still at an immature stage.

Prunus domestica

Rosaceae

Plum

Prunus domestica is known as the European plum, the domesticated plum, but also just the plum. It descended from the cherry plum, *Prunus cerasiforme*, either by hybridising with another plant in the *Prunus* genus or through evolution. The plum tree stands at around 9 m (30 ft) once fully grown and has serrated leaves that fall to the ground every autumn as it is deciduous.

The fruit of the plum tree is classified as a drupe, which is a fleshy fruit that contains a central stone surrounding the precious seed. The plum is just this, the seed being rarely seen by humans as it sits within the lignified endocarp of the stone, which was once the inner tissue of the ovary. The lignification helps to protect the seed. Surrounding this is the sweet, fleshy part known as the mesocarp, which is the bit that humans cultivate this tree for. The mesocarp was also once the inner part of the ovary. Keeping this all in check is the skin, which is known as the epidermis, and once again began life as the epidermis of the ovary.

Plums vary greatly in size, and shape, as they can be almost perfectly round or fairly elongated. They also come in a great range of colours – green, yellow, red, purple and almost black – depending on the amounts of the pigment anthocyanin found in the skin. The flesh is generally a creamy yellow.

The skin is also covered in a distinctive waxy layer, called the bloom. This layer helps to waterproof the fruit and stop water escaping. As the plum originated in the relatively warm climate of western Asia, this would have been a useful trait. To allow gases to move across the bloom, the skin is dotted with lenticels, spongy cells that can be seen as small freckles along the skin.

Plum

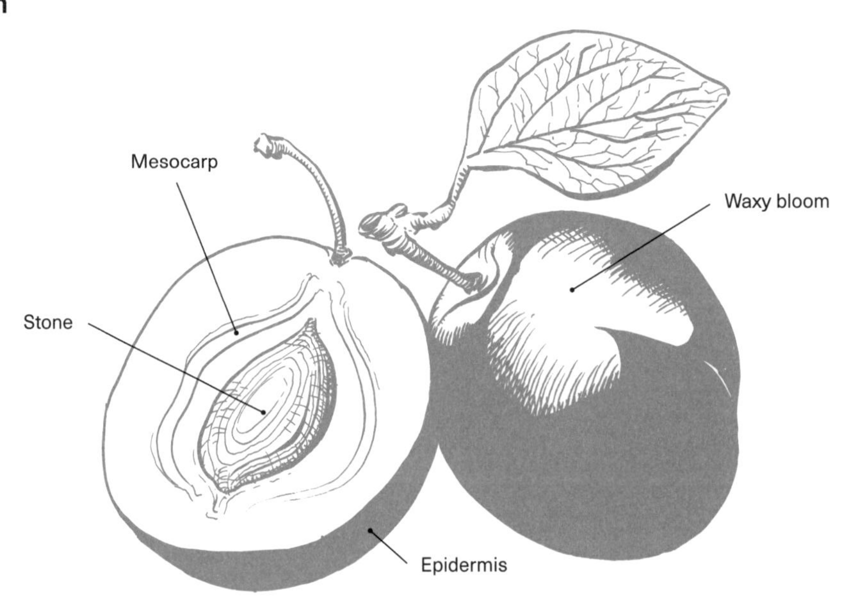

Almond

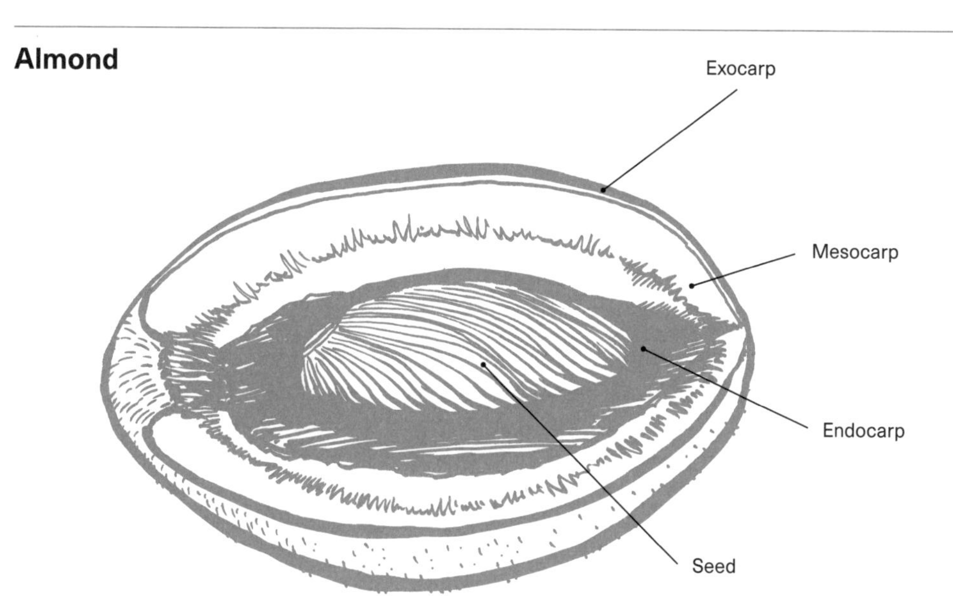

There is a subcategory of plums called greengages, which are usually green, but can also be yellow. They are relatively small and have a sweet but acidic flesh. They are also known as Reine Claude after the daughter of King Louis XII of France.

Another way of consuming a plum is to eat it dried, known as a prune. Not all plums will dry into a prune, as some have a tendency to ferment instead, so those with more solids and less juice dry more successfully.

Prunus dulcis

Rosaceae

Almond

Originating in Iran and the Tian Shan mountains, where many popular Rosaceae fruit crops originated, *Prunus dulcis* or the almond is a relatively small, deciduous tree that humans have cultivated for its edible seed. Although commonly classified as a nut in the supermarket, the almond is not a dry drupe, so does not qualify botanically.

There are two types of almond tree: the bitter almond, *P. dulcis* var. *amara*, and the sweet almond, *P. dulcis* var. *dulcis*. The fruit of the bitter almond contains high levels of amygdalin, which is broken down into cyanide gas, a toxin to humans, making the fruit inedible. Instead, bitter almonds are harvested for their oils, and during the extraction process the toxins are eliminated. Sweet almonds do not contain amygdalin and are eaten raw or cooked.

Both tree types need as much sun and heat as possible to produce a good crop, beginning after they are three years of age. The tree itself is hardy down to around –25°C (–13°F), but this temperature will destroy the fruit buds. The

The botany of composting

Most kitchen gardens contain a compost heap, somewhere to collect weeds, grass clippings and browning plant stems. In turn, this pile miraculously breaks down into rich, brown compost that can be spread on vegetable beds, feeding the soil and next year's crops.

What actually happens in the compost heap is a re-enactment of what occurs naturally when leaves and plant material fall onto the ground and are broken down by soil organisms. Plants are primary producers, making their own energy from the sun, and in turn these feed primary consumers such as humans, but also worms, woodlice and slugs to name a few.

What actually happens in the compost heap is a re-enactment of what occurs naturally when leaves and plant material fall onto the ground and are broken down by soil organisms.

Many of these primary consumers are saprophytes that feed exclusively on dead plant material, and these can be seen with the naked eye in the compost heap. There are often woodlice and brandling worms, *Eisenia fetida*, which can be observed living in the cooler parts of the compost heap, and which physically decompose larger components. These detritivores or decomposers eat the large pieces of plant material, and once this has traveled through their digestive tracts, is excreted in smaller pieces, ripe for the secondary consumers, the micro-organisms.

These smaller soil organisms feed on the waste of the primary consumers, breaking it down into yet smaller particles and releasing inorganic salts or nutrients in the process. These nutrients would otherwise be locked up in the dead plants, so when released are available to be used by living plants, once the compost has been spread on the soil.

Bacteria are the most common secondary consumer in a compost heap, along with fungi and actinomycetes. As the bacteria break down the compost, they release heat, which is why the centre of a heap is often very warm, although in home composting systems the centre is rarely warm enough to kill all weed seeds.

Fungi also break down the compost heap, but unlike most bacteria they can also decompose harder, woody tissues such as lignin.

Fungi also break down the compost heap, but unlike most bacteria they can also decompose harder, woody tissues such as lignin. Actinomycetes are bacteria, which act like fungi, producing mycelium and with the ability to break down some woody tissues.

Most of the secondary consumers need oxygen, heat and moisture to decompose the plant material, which is why a compost heap needs to be aerated and watered if it dries out. Aeration and mixing of the heap has two purposes, one being to give a supply of oxygen. If oxygen is not present then a different set of decomposers appears and breaks down the heap under anaerobic conditions, with one of the by-products of this being methane, creating a smelly compost.

Mixing the heap also mixes the nitrogen-rich 'green' components and the carbon 'brown' parts. The brown plant material is often much slower to break down, but the organisms that do this need nitrogen to operate. Once the nitrogen or green parts are exhausted, the browns also stop getting broken down, which is why it is important to have a good mix of the two.

tree needs 200–700 chilling hours to break dormancy, so requires a temperate climate to thrive. It produces its blossom early in spring. This unfortunately can be killed by a late frost, which is why almonds thrive in places such as California, but not the UK. Once the flowers bloom, they need to be cross-pollinated by insects, including bees, as the plant is usually self-sterile.

The fruit of the almond is classified as a drupe, similarly to other fruits in the Rosaceae family, and, despite appearances, is a fleshy drupe. Unlike other drupe fruits, the mesocarp or flesh is not what the almond is harvested for, as it is astringent and tough, so not a great ingredient. In the almond, the mesocarp is known as the hull, along with the downy, green exocarp or skin. The hull splits open when the fruit is mature, revealing the actual harvest.

What lies beneath the hull is the pit or stone, formed from the endocarp, which can be quite hard and lignified or thin and papery, depending on the type of almond tree. This hard stone is often botanically referred to as a pyrene and its appearance is a clue to the ancestry of the almond. The stone is slightly grooved, similarly to that of the peach and nectarine, all of which share a common ancestor. They are grouped in a subgenus called *Amygdalus* within the Rosaceae family.

Breaking away the stone reveals the seed, the part known to most as the almond. It has a brown seed coat, which is edible, but the almonds are often blanched. Blanching – soaking the almonds in hot water – softens and removes the seed coat, leaving just the bright white flesh of the seed.

Pyrus communis

Rosaceae

Pear

The pear tree often stands side by side with apples in a traditional orchard. In the wild, this deciduous tree reaches the heady heights of 20 m (65 ft), but is quite often kept at a more manageable height by grafting onto a dwarfing rootstock. The bark is dark brown and upon

Pear

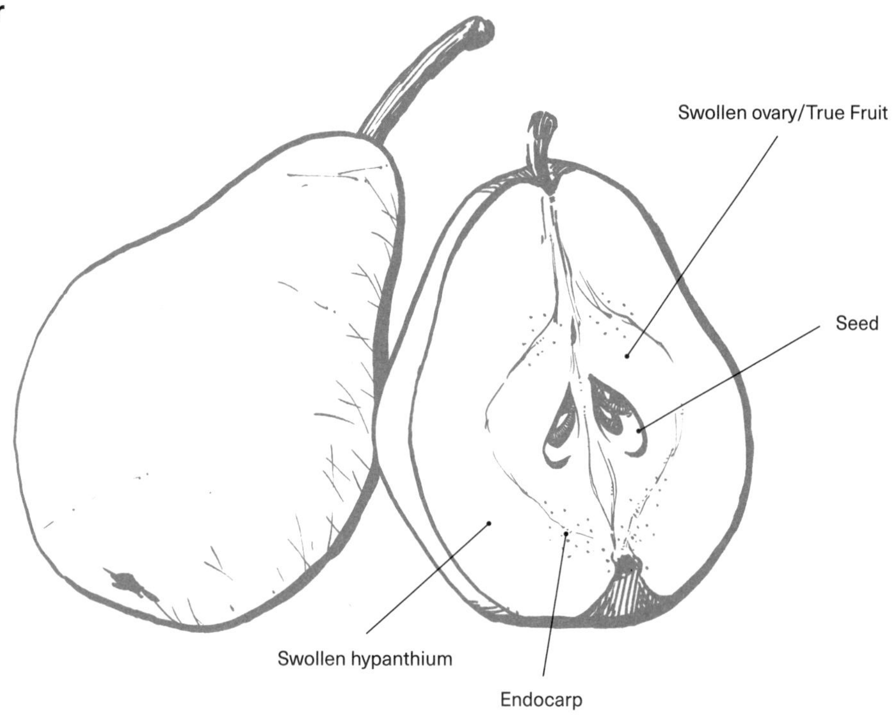

maturity breaks into squares, which is a ready identifier in the middle of winter in the orchard, as apple tree bark cracks into lines.

The fruit of the pear develops from an inferior ovary, which means it sits below the flower. This is obvious in the fruit as the flower parts are separated from the stalk or pedicel by the swollen ovary. The fruit is a pome fruit, as is an apple, and is botanically an accessory or false fruit. This is because the majority of the flesh is formed from the hypanthium, which is fused around the ovary to protect it. The ovary itself is within the swollen hypanthium, and the leathery endocarp forms the edge of the pear core that contains the ovary and seeds. The seeds sit within five locules or cartilaginous carpels that form the star shape when cut in a cross-section.

The flesh of the pear often dictates its use, as some varieties remain hard upon ripening and are used for cooking. Table or dessert pears have soft, buttery flesh upon ripening. Many pears have a gritty texture to their flesh, caused by stone cells. These gritty areas, also known as brachysclereid cells, are areas where the sclereid cells have undergone secondary thickening and filled themselves with lignin. Although it is unclear why pears have these cells, they may deter herbivores, and as cells are quite bitter, but humans have bred pears to have fewer of them, reducing this bitter flavour.

Pear skin varies from green to brown and red, but all have a slightly waxy texture, as there is a thin membrane to stop water loss from the fruit. The pear skin has very obvious freckles, known as lenticels. Lenticels are spongy cells that allow the movement of gases such as oxygen through the waxy membrane. These gases are used for cell function within the fruit.

Pyrus pyrifolia

Asian pear

The Asian pear is native to south China, and although it is in the genus *Pyrus*, in many ways resembles the apple, *Malus domestica*, more strongly in its characteristics. The fruit has a more rounded shape, unlike *P. communis*, which has the typical pyriform shape of a slim neck and fatter bottom. The flesh of the Asian pear is crisp, similar to that of the apple, but contains some stone cells. Asian pears can also be left on the tree to ripen, unlike *P. communis*, which is harvested before it is ripe, as it has a tendency to become very soft if left for too long.

Raphanus sativus

Brassicaceae

Radish, Mooli

The radish is an annual herbaceous crop, largely harvested for its immature, swollen root. It is rarely found in the wild now, but is quite likely to have originated in western Asia, with domestication beginning in Egypt by 2000 BCE. Some strains of radish are happy to grow in the higher temperatures found in these two locations, but it is generally seen as a cool-climate crop, thriving at around 10–25°C (50–70°F).

As with many crops described as root crops, this is not wholly accurate for the radish. The swollen appendage we consume is part primary root and part hypocotyl, the organ just below the seed leaves or cotyledons. The roots vary in shape from fusiform (long) to napiform (globular) and intermediate between the two. Harvested young, they are one of the fastest crops to develop from seed to being ready to be picked, but get quite woody with age. Although heat will make some bolt, any radish growing when the

Radish

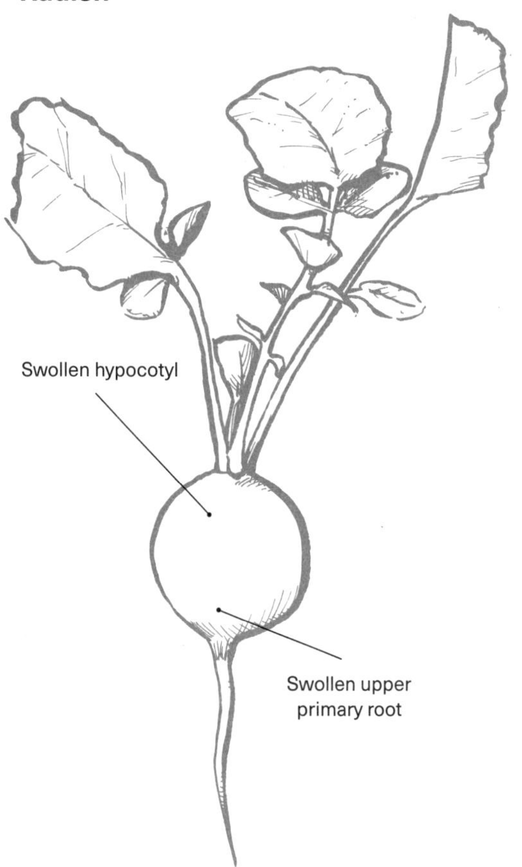

Rat's tail radish

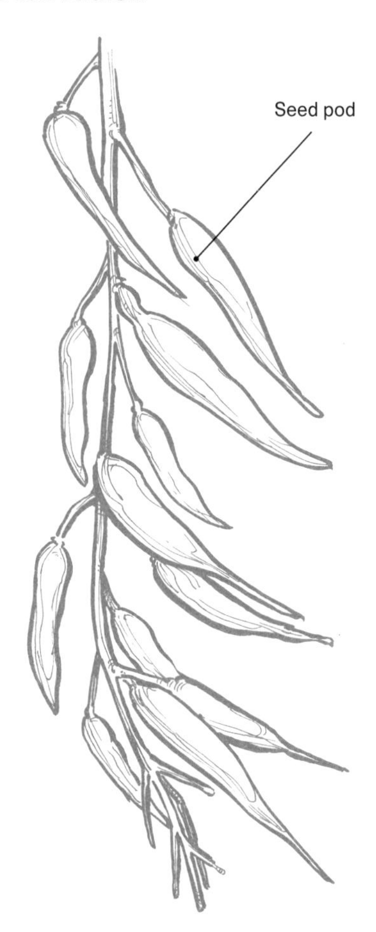

day length reaches 15 hours will go straight to seed, developing an oddly shaped root, because *Raphanus sativus* is a long-day plant.

The radish finds itself classified into several subcategories, mainly depending on use. Small or western radishes, *R. sativus* var. *ridicula*, tend to be red, red–white or white, and either spherical or long. They are harvested for their immature root and generally eaten raw. Oriental radishes, *R. sativus* var. *longipinnatus*, are also known as mooli. They have a milder flavour, but are much bigger in size than western radish and withstand the cold of winter well.

All radishes are sources of the so-called mustard oil bomb that gives them their strong flavour. As with many strong flavours, mustard oil is a deterrent to herbivores. The bomb aspect refers to the way two molecules are kept separate within the plant, until it is damaged by a predator, releasing the compounds to meet and create this strong-tasting compound. In this case, glucosinolate is broken down by the enzyme myrosinase. Much of the enzyme is in the skin, so peeling them can give rise to a milder crop.

The flowers on the radish have four petals, as is common throughout the Brassicaceae family. This is the origin of the previous family name, Cruciferae, meaning cross-bearing. The flowers are complete and entomophilous, being pollinated by insects, particularly the honey bee. They use cross-pollination, and stop selfing with a system called sporophytic incompatibility, where the growth of incompatible pollen is stopped and no pollen tube is created. This cross-pollination also extends to hybridising with *Brassica oleracea*, which leads to a series of crops called brassicoraphanus, used mainly for fodder.

Raphanus caudatus

Rat's tail radish

Another type of radish, sometimes classified as *R. sativus* var. *caudatus*, is known as the 'Rat's tail'. This has no swollen root and is instead cultivated and harvested for its crunchy, edible seed pods. Harvested when immature and green, anywhere from 5 to 20 cm (2–8 in) in length, the pods of the Brassicaceae family are known as siliques. They have two fused carpels, and the walls of the ovary separate when ripe – if they aren't harvested by a hungry human first.

Rheum x *hybridum*

Polygonaceae

Rhubarb

Rheum x *hybridum* is one of those amazing perennial crops that sits in many kitchen gardens. Known to most as rhubarb, it is a delightful spring treat that appears year after year with very little input from the gardener. As a native of riversides, it likes a cool season, and moisture, although it can survive drought periods. It also needs a cold period, 7–9 weeks at 3°C (37°F), to break dormancy and produce a healthy, thriving plant year on year.

As with many perennial plants, rhubarb emerges early in the season, and can be cropped all year, but it is important not to pick too thoroughly, which is why harvesting often stops in the summer. Over the summer and autumn, the plant can be left to develop and store food in its large underground rhizome, to ensure the plant produced in the following year is equally large and healthy. If continually picked, the plant cannot store enough food as it is constantly using reserves to put on new growth, and is likely to give smaller harvests the following year.

The part of the plant cooked for the plate is the leaf stem or petiole, which is the large, red, fleshy part that grows upwards from the ground, emerging from the buds in the crown. The petioles vary in colour from nearly green to pink to dark red, but the colour does not deepen as it ripens, and instead varies from cultivar to cultivar.

Thin, bright pink rhubarb has usually been forced, where light is taken away from the young shoots, making the plant use food reserves to search for the light and producing a sweet and tender crop. This practice does exhaust the rhizome of energy, so is not recommended on the same crown each year. Vascular bundles are strewn throughout the rhubarb petiole, and while these are barely noticeable at a young age, can be seen in older stems, as strands after the harvest has been completed.

Rhubarb is commonly eaten as a fruit, in sweet dishes, but as it is a vegetative part of the plant that is eaten, which has nothing to do with reproduction, it is in fact a vegetable. Early recipes often saw rhubarb in savoury dishes,

Rhubarb

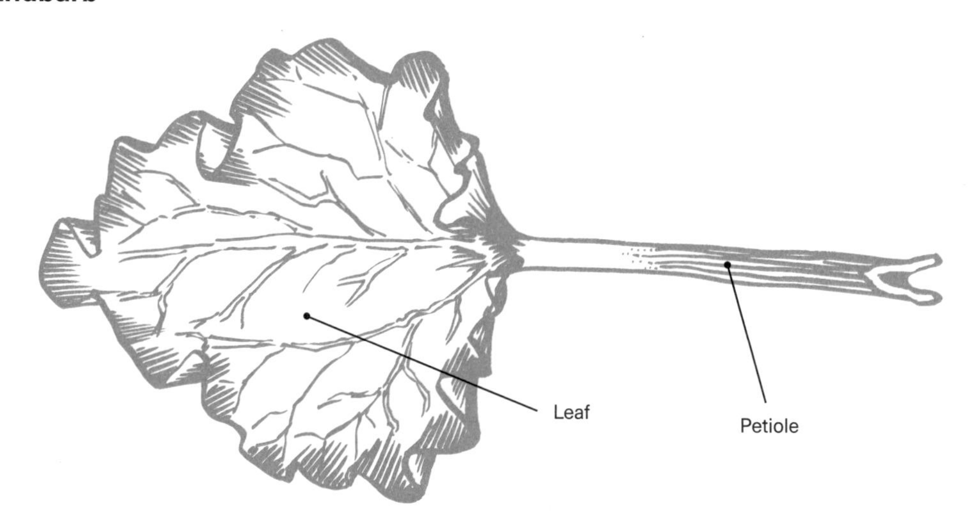

but more recently the fashion for tart flavours such as that produced by this crop has lessened. Instead, sugar is added to neutralise the acidity. Rhubarb is highly acidic, with the juice coming in at a pH of 3.2, mainly due to malic acid, but also a small amount of oxalic acid. The large, cordate leaves contain more oxalic acid, plus a toxin, anthraquinone glycoside, which can kill humans, so the leaves should be thrown away or composted, and never eaten.

Ribes spp.

Grossulariaceae

Currant family

Ribes is the only genus within the family Grossulariaceae and in the kitchen garden gives us deciduous shrubs that produce edible berries. A berry is formed from a single ovary, and has a fleshy pericarp but no stone. *Ribes* spp. have easily identifiable globed leaves, and small, almost insignificant, green flowers. The plants are monoecious, with imperfect flowers, both male and female on the same plant.

Although *Ribes* spp. can self-pollinate, fertilisation is seemingly better when cross-pollination occurs through insects. All three *Ribes* spp. commonly cultivated for fruits can hybridise, so rather than propagate from seed, they are reproduced via hardwood cuttings. The small shrubs are formed of canes, which are fruiting stems that arise directly from the ground.

Ribes nigrum

Blackcurrant

Blackcurrants are dark purple, almost black fruits, with the pigment anthocyanin solely in the skin, whereas the flesh remains green, which can be seen when it is cut into. Containing a fair few seeds, the flavour of this berry is often deemed tart with a sweet edge. The tartness comes from high levels of citric acid and ascorbic acid, or vitamin C, which made this fruit a great

Blackcurrant

replacement for citrus at points in history when the latter was unavailable.

Native to central and northern Europe and northern Asia, all *Ribes* shrubs need a period of chilling, 1,000 hours in the case of the blackcurrant, to break dormancy in spring. The buds are actually formed in the previous summer, at which stage they are undifferentiated. Triggered by autumnal day length, the buds differentiate into either a flower or vegetative bud, and with blackcurrants, generally those towards the base of the cane all become flower buds, which is why when pruning, we leave one- and two-year growth, but remove the entire limb after this, as the majority of the buds will differentiate into vegetative buds. The buds that differentiate in the autumn will break the next spring, and as long as they receive frost-free days of above 5°C (41°F), will bear fruit around 120 days later.

The leaves of the blackcurrant are particularly aromatic and contain glands on the underside that help with pest and disease defence. The aroma comes from thiol compounds, often known as 'cat ketone' as they are heavily present in cat urine, and many people smell cat when they sniff blackcurrant.

Although individually insignificant, the flowers of the blackcurrant hang in racemes of around 10–12, with a raceme arrangement whorling around a central stem on short stems, at equal distance from each other. The observant will see it is always the lower flowers that burst first.

Ribes rubrum

Redcurrant, Whitecurrant

Although visually very different fruits, redcurrants and whitecurrants are botanically the same species. The stark difference in the colour of the fruits results from the lack of anthocyanin pigments in the whitecurrants, which are essentially an albino redcurrant. Both berries have a translucent skin, and the redcurrant has

all its pigment in its flesh. The berries tend to be smaller than those of the blackcurrant, but also contain less seeds.

The flowers form racemes, usually of around 14 flowers; this grouping in edible *Ribes* is known as a strig of fruit. The buds of *Ribes rubrum* are most commonly formed on tiny spurs that emerge from the central cane. The buds are formed on one-year spurs, but quite often these spurs emerge from older wood, which is why the canes are often allowed to live for four years before being cut down to the ground in the dormant season.

Ribes uva-crispa

Gooseberry

Gooseberries are much bigger than the black, red and white currants, and come in several colours, yellow, green and red. They are often slightly hairy, a trait that has been bred out of some cultivars. The plant has less apical dominancy than blackcurrants and redcurrants, which means it is often less of an upstanding plant, and instead branches and spreads sideways. The canes are covered in nodal spines, from modified leaves, which keep large herbivores away from the fruit. The flowers of the gooseberry are usually singular and do not form racemes like the other Ribes species.

Rubus idaeus

Rosaceae

Raspberry

The raspberry, *Rubus idaeus*, is a caneberry that grows wild in Europe and western Asia. It is a perennial, deciduous shrub, which is hardy down to extreme colds of –20°C (–4°F). The stems of the raspberry are known as canes, as they originate straight from the ground and are hollow with pithy layers. Canes are extremely strong

The botany of storing harvests

Harvests such as squash and sweet potatoes are cured before storing. This ensures the skin is dried and watertight, allowing any wounds to heal over, known as suberisation.

Although many crops are best eaten fresh, some can be stored for several months, a particularly useful trait in the middle of cold seasons when little is growing in the kitchen garden. Even after being harvested, the crops continue to live and therefore have metabolic activity that needs to be controlled to prolong the storage period.

Many crops require a lower water content to store well, such as beans and some root crops. A higher moisture content allows fungal spores and bacteria to reproduce, encouraging the crops to rot in storage. Drying a crop fully is fairly straightforward: leave it in a warm, dry, well-ventilated space, ideally still in the kitchen garden or on the plant, if the weather allows. The warmth encourages the movement of water, the dry conditions stop the reuptake of water and the ventilation moves away any humidity, keeping the concentration of water outside the crop lower than inside.

Often it is simply enough to dry out the skin or rind of the plant fully to stop moulds growing. Harvests such as squash and sweet potatoes are cured before storing. This ensures the skin is dried and watertight, allowing any wounds to heal over, known as suberisation. Suberisation is when the cells are filled with suberin, a cork-like, watertight material.

Curing is undertaken in similar conditions to drying, but often at slightly warmer temperatures, which are crop-specific. Because the plant tissue is still alive but can no longer photosynthesise, some of its stored starches are converted into sugars for respiration to provide energy, which has the added benefit of sweetening the crop. To slow respiration and ensure not all the sugars are used up in respiration, both dried and cured crops are kept in dark and cool environments. Ventilation is also important to remove a build-up of carbon dioxide from around the crop, which is produced during respiration, as this depletes oxygen from around the cells and can lead to fermentation.

Climacteric fruits continue ripening after harvest, so it is best to keep them in a cool, dark area to slow respiration.

Most fruits need to be ripe to store well. Ripening is controlled by a plant growth regulator called ethene, which is one of the few regulators that is a gas. Some crops experience a sudden ripening, where they use their reserves in a burst of respiration, called the climacteric rise. Climacteric fruits continue ripening after harvest, so it is best to keep them in a cool, dark area to slow respiration. These fruits produce ethene during the climacteric rise, which in turn can stimulate ripening in nearby fruits, so is a good idea to harvest them just before they are fully ripe to allow this process to happen in storage.

Non-climacteric fruits such as citrus and cherries do not continue to ripen after harvest, nor do they experience a sudden rise in respiration, but will slowly soften with time. There are some crops, such as carrots and cabbage, that although they do not produce great amounts of ethene, are quite sensitive to it, and it can decrease their storage life. So it may be wise to separate these from the climacteric harvests in storage.

and can grow to heights of 2 m (6.5 ft) with no support; other cane plants, such as bamboo, can grow even higher.

The flowers on the cane are hermaphrodite or perfect flowers and contain both male and female organs. They are self-fertile, but cross-pollination by insects such as bees increases pollination and fruit production. *Rubus* spp. can produce seeds via apoxis, which is asexual reproduction where a clone is created from tissues of the ovule.

Not berries at all, the fruits of the raspberry shrub are in actual fact aggregate drupes or drupelets. The fruit is fairly obviously made of up of lots of different fruits, as seen by the bumpy nature of the skin, all from separate ovaries contained within the flower. These fruits are fleshy and contain a stony endocarp, which in turn surrounds the seed, making them drupes, similarly to stone fruits such as peaches and plums.

These drupelets are attached to a tiny core, which is formed from the receptacle. The core, sometimes known as the torus, is dry and fleshy and unappealing, but thankfully in raspberries comes away from the fleshy drupelets and is often left on the cane, causing a hollow in the centre of the tiny fruits. Frustratingly, the receptacle remains in other caneberries when they are picked, such as blackberry, *R. fruticosus*, and can give a drying texture.

The fruits are generally pink to red in colour owing to the presence of anthocyanins, but there are also yellow or gold types, which are much more rare. These yellow fruits have a recessive gene that restricts the amount of anthocyanin produced, resulting in the golden flesh. All raspberry fruits are a little hairy in texture, but these are not true hairs only the remains of the pistils.

Raspberries are grouped into two growing types, summer and autumn. Autumn types fruit

Raspberry

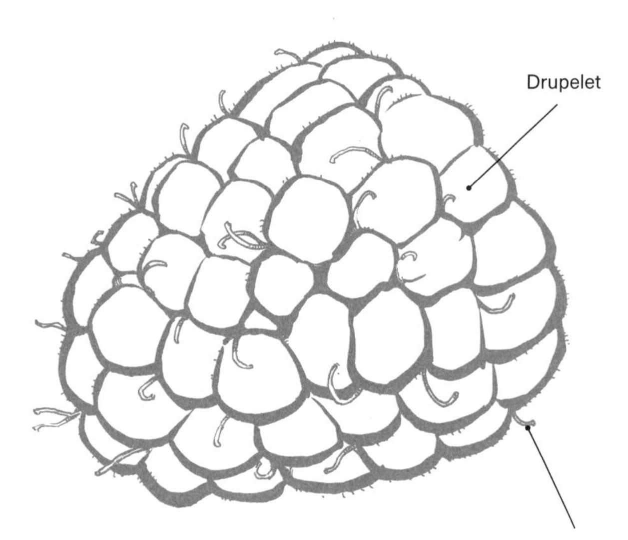

on one-year-old canes, known as primocanes, and once fruited, are cut down to the ground to allow for new growth the following spring. Summer raspberries fruit in their second year on floricanes, or two-year growth. In the wild, most raspberries are summer raspberries and are biennials. They produce primocanes in their first year that simply grow upwards and give the height. Then in the second year, the now floricanes branch out and produce flowers. This is the end of the life cycle for these canes, but the perennial roots are constantly spreading out and putting up new stems.

The canes of raspberries and other caneberries, such as blackberries, *R. fruticosus*, are covered in tiny prickles as a defence against hungry herbivores. These prickles are sharp protrusions from the skin of the cane and are simply cellular divisions of the epidermal cells. They contain no vascular bundles, so are fairly easy to remove from the canes, and some cane fruits are now bred to be prickle-less.

Rather confusingly, these sharp prickles are often referred to as thorns, but botanically, a thorn is a modified branch, and a spine is a modified leaf, both of which contain vascular bundles, making them much harder to avoid or remove.

Solanum lycopersicum

Solanaceae

Tomato

Originating in the warm climes of South and Central America, the tomato was domesticated in Mexico and the Andes thousands of years ago. A member of the nightshade family, it was received with suspicion in Europe, owing to its resemblance to poisonous plants such as deadly nightshade, *Atropa bella-donna*, but finally the sweet-tasting fruits of *Solanum lycopersicum* became a staple of the table.

Older texts refer to the tomato as *Lycopersicum esculentum*, which is the synonym of the now permanent name of *S. lycopersicum*. There are five possible varieties:

- var. *commune* – common tomatoes
- var. *cerasiforme* – cherry tomatoes
- var. *grandiflorum* – large-leaved tomatoes
- var. *validum* – dwarf, upright tomatoes
- var. *pyriforme* – pear-shaped tomatoes.

Tomatoes are mainly self-pollinated, which means the seed they produce will come true to type when planted. In the wild these flowers are pollinated by the halictid bee, but these don't live in all the places humans have chosen to grow tomatoes, so self-pollination is a useful trait. Self-pollination is possible because the flowers of the tomato are complete, but the stigma does hang beyond the anther, allowing for cross-pollination to occur, usually around 14–30% of the time.

The role of bees is for buzz pollination. The anthers of the tomato are fused together, forming a hollow tube where the pollen is produced, but the pollen is only shed when moved by the correct vibrations. Bumble bees and carpenter bees vibrate at the right frequency, but sadly honey bees do not.

The tomato itself is botanically a fruit, a ripened ovary, but it is still OK to have them on top of a pizza. Classified as a berry as it is produced from a single ovary, it is fleshy and contains seeds, with the green calyx sitting in the top where the peduncle joins. Inside the fruit are the locular chambers, generally two in a cherry tomato, three in a common tomato and multiple locules in a beefsteak.

The seeds within the chambers are suspended in a mucilaginous substance known as the locular gel. This is an important substance, which starts life as placental cells. It produces ethene, which starts the ripening, so to all intents and purposes, although the tomato appears to

Tomato

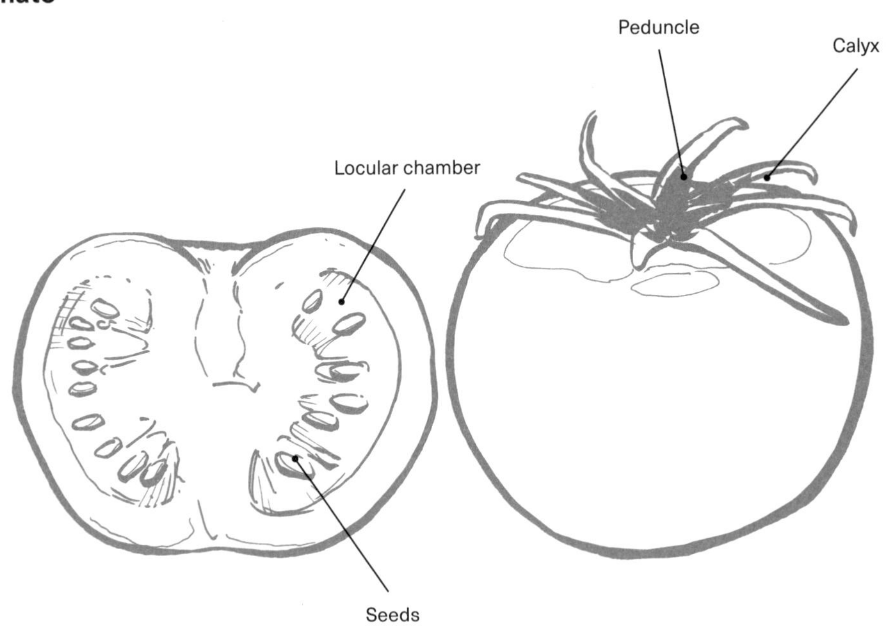

ripen from the outside, the whole process starts at the centre of the fruit. The gel is also involved in flowering, seed development and dormancy.

Although synonymous with red, tomatoes come in a variety of colours, and the hue perceived by humans is a combination of the skin colour and flesh colour combined. For example,

- Red: red flesh, yellow skin
- Pink: red flesh, clear skin
- Purple: dark red flesh, clear skin, high proportions of chlorophyll, particularly in the seed gel
- Brown: dark red flesh, yellow skin, high proportions of chlorophyll, particularly in the seed gel
- Yellow: yellow flesh, yellow skin
- White: white flesh, clear skin
- Green: green flesh, clear/yellow skin.

Originally, many tomatoes were naturally red with green shoulders, but the mass market decided this looked unripe and the trait was bred out of them to produce a uniform skin colour. Unfortunately, the gene that was turned off to express this uniformity, known as *Golden 2-Like* (*GLK-2*), also affects sweetness in the fruits. When turned on, it increases photosynthesis, thereby increasing sugar production, so by turning it off for the perfect-looking fruit, we have created a less sweet harvest.

The tomato plant has two distinct growth habits: indeterminate and determinate. Indeterminate cultivars are perennial in warmer climates, with their terminal buds always vegetative, and they produce fruit on trusses along their stems gradually throughout the season. These sprawl and are known as vines. When trained in the kitchen garden, the vines

are attached to stakes and their side shoots pinched out to stop them putting excessive energy into vegetative growth, and more energy into producing fruit. This is especially important in areas that experience frost, meaning the tomato has a shorter lifespan than it expects in the wild, and cannot grow perennially. This style of training on a single stem means indeterminate tomatoes are also known as cordons.

The determinate tomato plants grow upwards to a certain height and then stop, making them annual. All terminal buds produce flowers, which turn into fruits that generally all ripen at once, and the number of flowers produced is limited. In cultivation, these are called bush tomatoes. Breeders have now developed semi-determinate types, which grow as a bush, meaning less staking for growers, but produce fruit on trusses, meaning an extended season of harvest.

The leaves and stems of the tomato plant are pubescent, being covered with tiny trichomes (hairy appendages), which have many uses. When these trichomes hit the ground, they produce roots, giving the plant an extra source of water, nutrients and stability. These hairs also protect the plant from pests and weather.

There are two types of trichome involved in pest defence: glandular, which are round in shape and burst when touched by a predator, releasing defence proteins that trap or poison the insect (this is what causes gardeners' fingers to go green when sideshooting);

and non-glandular, which can sense insects walking and send signals to the plant to prepare for attack. The non-glandular type also provide a layer to stop raindrops causing damage to the leaf cells, holding the water above the epidermis.

The final aspect to take a look at on the *S. lycopersicum* plant is the leaves. While most commonly they are rugose with serrated edges, some have recessive genes, which cause a 'potato leaf'. Potatoes are close relative (so close in fact, they can be fused together to grow a potato underground and tomato above). Another leaf type is the angora, covered in tiny hairs, giving a silver colour.

Solanum pimpinellifolium

The currant tomato

Originating in the low elevations of Peru is another species in the *Solanum* genus, which produces tiny tomato fruits, hence the common name of currant tomato. There are some cultivated varieties, producing large amounts of these little fruits, but the real star of *S. pimpinellifolium* are its genes. One of the worst diseases to hit the tomato is late or potato blight, *Phytophthora infestans*, which destroys crops overnight, turning them into a slimy, stinky mush. The currant tomato is blight-resistant, and this trait is commonly bred into *S. lycopersicum* to help it with the continuing battle against this devastating disease.

Solanum tuberosum

Solanaceae

Potato

The ever-demeaned 'humble potato' is in actual fact an exotic traveller, having reached the UK from the Andes by way of Chile and Spain, and an extremely significant crop. It is one of the five most important crops grown in the world, feeding many millions. Potatoes contain a lot of protein in the form of the storage protein patatin, and the average potato provides 45% of the recommended dietary allowance (RDA) of vitamin C. With optimum growth occurring at 16–20°C (60–68°F), slowing above this and stopping completely at 29°C (84°F), potatoes are well suited to the cool spring weather in the British Isles and thrive here.

The potato is a perennial, herbaceous plant, but its aerial stems grow in an annual fashion, dying back each year, with new stems growing up from new parts of the underground storage unit. As it is the underground potatoes that are harvested annually, they are treated as annuals in the kitchen garden. Most parts of the potato that are left in the soil over winter cannot withstand the cold and wet, and rot off, but some parts do remain and occasionally send up shoots in the spring; these are known as volunteer potatoes.

The potato is often categorised as a root crop as it grows deep underground, but the tuber that is consumed is actually a swollen stolon or underground stem. The stolon is adapted to become a storage unit that holds essential nutrients, water and high levels of starch that the spud can use in times of drought, which happen fairly often in its native South American home. Starch is the common storage unit of energy for many plants, as it is extremely compact. In perennial plants, this starch is stored until the plant requires it – generally the spring, when it breaks down to sugars, and releases its energy.

On the surface of the potato skin are lots of indentations, commonly referred to as eyes. These eyes are nodes that contain buds that could become underground stems bursting through and eventually forming new adventitious roots, and stolons, which end in new, tasty potato tubers. The end of the potato where the stolon attaches to the tuber is known as the proximal end, and the opposite is the distal end, where the eyes are closer together.

Potato

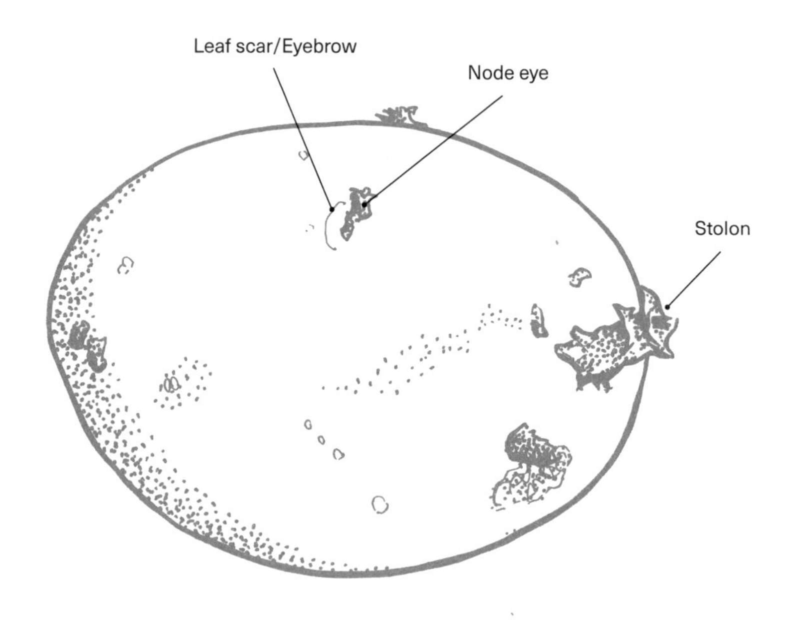

As the eye is botanically known as a node, the gap between them is an internode.

This terminology is more commonly associated with leaves, and the nodes are arranged as leaves around a stem, in a spiral pattern. Around this eye an eyebrow can usually be observed, this being the scar from a leaf scale. The stolon that is produced travels underground, forming new swollen storage areas, which become the potatoes we eat that season.

Lenticels cover the skin of the potato, allowing gas exchange in the tuber. If there is not enough oxygen, for example if the soil is waterlogged, the lenticels will expand, causing popcorn-like spots on the surface of the tuber.

The potato was originally an Andean crop, growing near Lake Titicaca, but the tubers growing there differ quite significantly from those cultivated. Originally a bitter tuber that contained poisonous alkaloids, they were eaten with clays, which would bind the poisons, rendering the tubers edible.

Another stark difference between modern and ancient potatoes is when the plant starts tuberising. The wild potato was a short-day plant, meaning it started saving up starch when the day length became shorter than the night time, which makes a lot of sense when growing in the Andes. In Europe this trait meant that the tubers started being produced in early autumn, but as this crop is not frost-hardy, the foliage was often killed off in the frosts, leaving small, insignificant potatoes.

The wild, Andean short-day potatoes are now thought to have been a subspecies, *andigena*, which is a diploid plant with very long stolons, and a natural hybrid of *S. stenotomum* and the weed species *S. sparsipilum*. These were the potatoes introduced to Spain in the 1570s. The potato more commonly grown today was

Blueberry

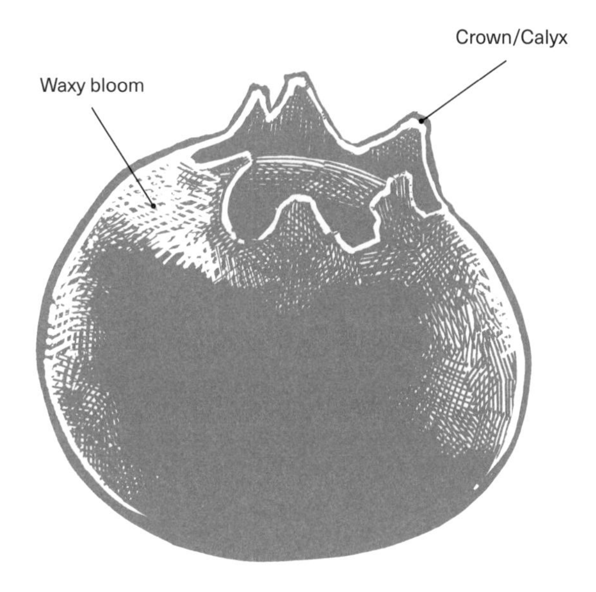

brought to the UK by the way of the USA in the 1800s and originates on the coast of Chile. It is descended from *S. tuberosum* subsp. *andigena*, also known as *S. tuberosum* subsp. *tuberosum*. It is a tetraploid plant with a long day length, and therefore more suited to cultivation further away from the equator. Slowly, this trait has been bred out and *Solanum tuberosum* is now classified as a long-day plant or day-neutral.

Potatoes are classed as waxy or floury, depending on the amount of dry content in the flesh. Waxy potatoes have a low amount of dry, starchy content, and are higher in sugar and moisture. These hold their shape well when cooked, and are commonly used for boiling and roasting. Floury potatoes have higher dry content, and less sugar and moisture, so do not hold their shape when cooking, making them perfect for mashing and frying. Floury types store more successfully owing to their low moisture content.

Vaccinium corymbosum

Ericaceae

Blueberry

The blueberry is a small, deciduous shrub that flourishes in an acidic site, around a pH 5, although it will happily grow in more neutral soils as well. Plants that prefer acidic growing situations are known as ericaceous, and mainly appear in the family Ericaceae, alongside *Vaccinium corymbosum*, the blueberry. This adaptation to low pH relates to the family's wild origins in boggy wetlands, where the water contains large amounts of decaying matter and has low oxygen levels, which create an acidic environment.

Another adaptation, seen in many ericaceous plants, is a lack of roots hairs. The root system is shallow and fibrous, but to increase the nutrient uptake, the roots have symbiotic relationships with mycorrhizal fungi. These take some sugars from the plant, while vastly increasing the surface area that the plant has to find nutrients, water and oxygen as the fungi have an extremely large network of strands called hyphae. In boggy conditions, there is very little in the way of available nutrients and oxygen, so this relationship is critical to the blueberry.

The flowers of the blueberry are monoecious, with separate males and females on the same plant, and like many other ericaceous blooms they are bell-shaped. This shape fully encloses the sexual organs, but makes self-pollination very tricky as the large, sticky pollen cannot escape easily. Instead, the flowers rely on insects, particularly bees, for pollination.

To stop the pollen simply falling out of these upturned flowers, they employ a system of buzz pollination. The visiting bee vibrates at a certain frequency, a process known as sonication, which in turn triggers the release of the pollen from the anthers. Not all bees vibrate, including honeybees, which instead transfer pollen less successfully by accidentally rubbing up against loose grains when feeding on the nectar.

Blueberry fruits are true berries, and are virtually blue in colour, so for once the common name is botanically correct. A berry is a fleshy fruit that has no stone and is produced from one flower with a single ovary, which nicely sums up the blueberry, with its blue skin and translucent flesh surrounding the seeds. One distinctive feature of the blueberry is the flared crown on its base, which is formed from the calyx.

The blueberry starts life as a small, green fruit, and then ripens through a deep red colour to the blue it is synonymous with, owing to anthocyanin pigments. There are some pink-skinned varieties of blueberry, which have differing amounts of anthocyanin. The skin also has a waxy bloom that lends the appearance of dusky patches to the fruit. This bloom stops water loss through the skin of the blueberry.

Vicia faba

Fabaceae

Broad bean, Faba bean, Fava bean

This bean goes by many common names and in the UK is one of the first spring harvests in the kitchen garden from overwintered plants. The broad bean, *Vicia faba*, is an annual, herbaceous plant that can grow up to 1 m (3 ft), but there are dwarf cultivars that only reach 30 cm (11 in). It is a cool-climate crop, with ideal growing temperatures of around 16°C (60°F), and some cultivars will happily flourish through the colds of winter to produce a particularly early harvest.

Starting life as a seed, broad beans are known as hypogeal plants, which means their cotyledons, or seed leaves, which can be seen when the seed is cut in half, remain underground after germination. In opposition to this are epigeal plants, whose cotyledons emerge above the ground and begin to photosynthesise. The cotyledons of the broad bean give food to the emerging plant, but shrivel and die beneath the soil, and instead it is the plumule, or initial stem, that breaks the surface and produces true leaves.

Hypogeal plants benefit as their seeds can survive being deeply buried, up to 25 cm (10 in) for the broad bean; additionally, if there is any damage to the young growth above ground, from frost or pests, the plant can grow again from underground nodes. By contrast damage to the above-ground cotyledons in an epigeal plant will generally result in the death of the plant.

Broad beans have one central stem, which is recognisable from its squared shape, and does not send out any lateral branches. It will put up new, smaller stems from the nodes at the base of the stem, known as tillers. If any damage occurs to overwintering stems, the plant is likely to tiller and put up new, but smaller stems.

At the bottom of the leaves of the broad bean are small green bracts, which sometimes have a

Broad bean

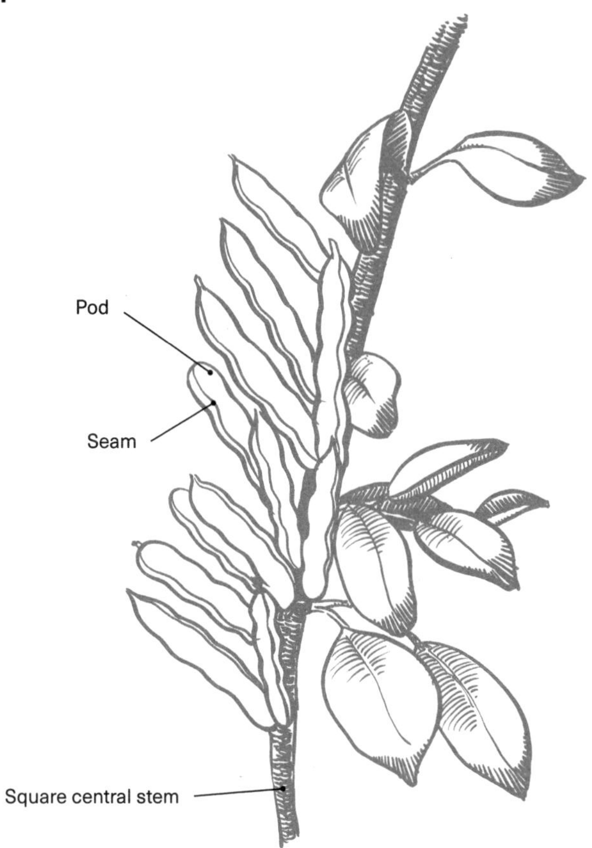

dark patch, known as the stipule. These stipules contain nectaries, which are normally found in the flower to attract pollinators. Although it is not certain why broad beans put energy into creating these stipules, one theory is that it is to attract ants, which will protect the plant from caterpillar attacks.

The flowers of the broad bean are recognisably those of a legume, with keels and wings. The broad bean produces excessive numbers of flowers, with the expectation that only around one in five will set to become fruits. This is an evolutionary tactic to attract pollinators and ensure fruit set.

Broad bean fruits are legumes, as they are dry and dehiscent, splitting along two seams when ripe. The pods grow upwards in the leaf axils in small clusters, and can be harvested either when immature, or dry and mature for their seeds. The pod has a spongy, slightly hairy endocarp, and the inside of the pod is white and downy, giving extra protection to the seeds.

Long-pod broad bean pods will contain 5–8 seeds, whereas anything with four or fewer is classified as a Windsor broad bean. The seeds have a tough testa or seed coat, which becomes tougher with maturity, and many people remove this after cooking to simply eat the fleshy seed. The beans are high in protein, as are many legumes, because the plant has a symbiotic relationship with nitrogen-fixing bacteria, which allows the plant to store food as protein.

Vitis vinifera

Vitaceae

Grape

Vitis vinifera is a woody climber, which if left unattended would attain the great height of 20 m (66 ft). But these vines, known to most as grapes, are predominantly found in cultivation where they are pruned to more modest heights and lengths. Keeping the growth of the grape young has other advantages, as after two years, the fruit produced is smaller and of poor quality, so the aim in cultivation is to keep producing new, young shoots. The woody, lignified trunk of the grapevine has a beautiful bark that has deep crevices and peels away as the oldest bark dies.

The vine climbs over other plants and objects to reach the sun, and uses forked tendrils made from modified stems to do so, which in the case of the grape are made from modified stems. The new stems of the grape appear from buds, which are known as compound buds, as they are grouped in threes. The centre bud is always the first to break, and always receives the most energy – in fact the other two don't seem to do very much, unless this primary shoot is damaged, in which case one of the other buds will grow away and take its place.

The leaves of the grape are large, palmate and lobed, and although generally green, can be tinged red as well. The leaves are commonly used for food in some cultures, such as making the Greek dish dolmades. The leaves grow along the new shoots of the vine, as do the flowers, which are hermaphrodite and hang in panicles or loose clusters, which will eventually turn into a bunch of grapes. A bunch of grapes consists of the berries, but also the rachis, which is the stem in the centre of the inflorescence.

The fruit of the grape is a berry, as the endocarp and mesocarp are both fleshy, and do not contain a stone, while the epicarp forms the skin. These berries tend to contain locules and around four seeds, which can be seen inside the grape. There are also seedless grapes, known botanically as a pyrene, and although marketed as having no seeds at all they are still present but are just so soft and indistinct that they aren't noticeable.

The skin of the grape varies greatly in colour from nearly white, through green, red and purple to nearly black. The difference in colour is controlled by a gene that dictates the amount of anthocyanin produced. The epicarp or skin of the grape is covered in a waxy bloom, like many fruits, which helps reduce water loss in the arid conditions the grape thrives in. This bloom is also the home for natural yeasts, in this case *Saccharomyces cerevisiae* var. *ellipsoideus*. These yeasts contain the enzymes that break down grape sugars and allow fermentation, resulting in wine.

Grapes are loosely categorised by how they are eventually presented to the consumer, as either table, wine or dried. Table or dessert grapes have a firm flesh, thin skin and a low acidity, but are not as sweet as wine grapes. Although most table grapes are *V. vinifera*, some come from the Muscadine grape, *V. rotundifolia*, with its nearly rounded leaves, and the fox grape, *V. labrusca.* The fruits of the fox grape are sometimes called slipskin, as their epicarp parts very easily from the mesocarp flesh.

Grape

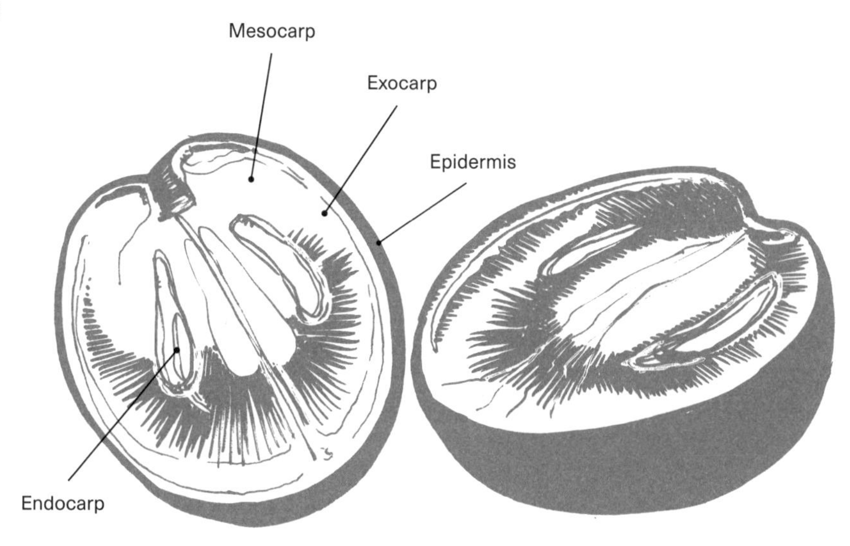

Wine grapes are relatively smaller than table grapes, and have a softer flesh, but are much sweeter with a thicker skin. The higher sugars help produce the alcohol in wine, and the thick skins impart the colour of the wine, which is particularly important in red wines.

Finally, there are grapes for drying. Once dried, the fruits are called currants, raisins or sultanas. Currants come from small, black, seedless grapes and are the smallest of the dried fruit, with maybe the strongest flavour. Raisins are any grape, and are dried for around three weeks, making them dark in colour. Finally, sultanas are green, seedless grapes, which are coated in oil before drying, which means they do not need to be dried for as long and often are lighter in colour and plumper in texture.

Zea mays

Poaceae

Corn, Maize, Sweetcorn

Zea mays originates from the warm climates of South America and was probably domesticated in Mexico. It is a monocot, meaning it only has a single cotyledon upon emergence, and features strap leaves with parallel veins, with their vascular bundles scattered throughout the tissues, rather than in a ring. Commonly known as corn or maize, it is a herbaceous, usually single-stemmed annual grass, which humans cultivate for its edible seeds. The seeds can be eaten when immature or dried, at which point this crop is known as a cereal, a grassy crop harvested for its starchy grains. The large, single stem or culm of the corn plant can grow to 3 m (10 ft), and even higher in the wild.

There are three types of roots connected to the base of the corn plant. The seminal roots are produced first and include the radicle, or seed root, and around three other initial roots, which become inconsequential once the plant matures. Roots produced from an underground node are known as the coronal roots, and those produced above ground are buttress or brace roots that develop once the plant starts to grow tassels. The buttress roots support the plant, especially if it is growing in an open environment or shallow soils. They have the added advantage of the aerial parts being photosynthetic until they reach the soil.

Corn photosynthesises slightly differently from many other plants, using the C4 metabolic pathway. This is fairly common in plants that grow in hotter climates as it reduces energy loss compared with standard photosynthesis, sometimes referred to as C3 photosynthesis. In addition, the C4 metabolic pathways almost halve the amounts of water needed for C3 photosynthesis, which is extremely advantageous in drought conditions.

The leaves of corn are linear and lanceolate (lance-shaped), with a hairy upper margin. As well as growing from the stem, they also develop around the fruit or cob. These leaves are more enlarged and overlap to create a sturdy protective barrier, known as the husk or shuck. Sweetcorn is often sold with the husk still attached, and this further protects the crop in transit.

The flowers of the corn plant are monoecious, with the male or staminate flowers growing at the top of the stem, and the female or pistillate flowers growing in the axils of the nodes. The male flowers are collectively known as the tassel and have a compact, branched, panicle formation. As corn is wind-pollinated, growing at the tip of the plant allows the pollen to be easily caught on the breeze.

The female flowers develop at the end of a short, specialist stem, known as the shank, and have short internodes. The female flowers have extremely long styles, known as silks, and are what can be seen as the end of a corn cob. These often need to be removed when eating sweetcorn. Unlike most styles, which have a sticky stigma at the end to catch the pollen, the silks have sticky trichomes all along them, increasing the chances of catching pollen, and undergoing pollination.

Once pollinated, the cob develops rows of around 300 seeds known as the kernels. The kernel or fruit is botanically classified as a caryopsis, as it is dry, with one seed, and indehiscent. This is similar to the description of an achene, except with a caryopsis, the pericarp is fused with the seed coat and forms a strong protective layer. The hull of the kernel is made up of the pericarp and seed coat, and the germ is formed from the endosperm and embryo, so the majority of what is eaten is the endocarp. These kernels form around a cob, a tough cylindrical structure, which is harvested for corn on the cob. Corn varies widely in colour from yellow, to black, blue, red, green and white.

Occasionally corn is harvested before pollination occurs, when it is known as baby corn. As well as baby corn, there are six main classifications of corn:

- Dent corn has both hard and soft endosperm, but as the soft endosperm contracts, it leaves a distinctive dent in the top of the kernel, giving it its name. It produces corn meal and corn syrup.
- Flint corn has a very hard hull and is often brightly coloured with reds and purples, making it perfect for ornamental decoration.
- Pop corn is a type of flint corn with a starchy endosperm, and is used to create the snack popcorn.
- Flour corn has a soft endosperm and thin hull, making it perfect to grind down for flour.

Sweetcorn

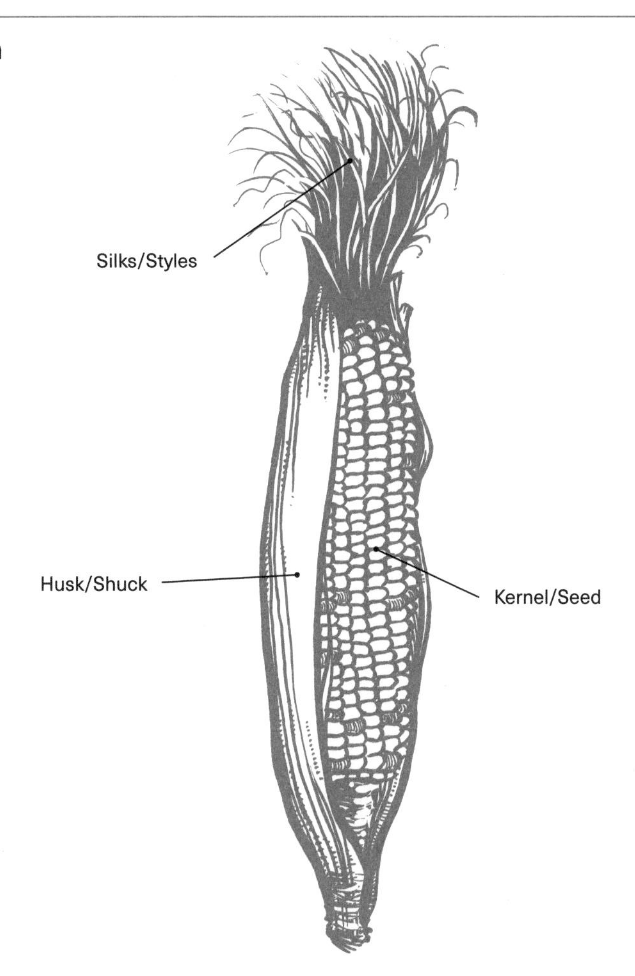

- Pod corn is wild and has a husk around each individual kernel.
- Sweet corn is the type most commonly associated with the kitchen garden. It is harvested at an immature stage, and the endosperm is high in sugar.

Corn has undergone much breeding to improve its sweetness. 'Sugary' or 'su' corn has the *su* gene, which blocks the conversion of sugar to starch once it passes from the leaves to the kernels, but when harvested, these sugars do revert to starch fairly quickly. 'Sugary enhanced' or 'se' corn slows the post-harvest conversion of sugar to starch, but has such tender kernels it is hard to mechanically harvest. And finally, 'supersweet' or 'sh2' hybrids have twice the sugar of the standard corn, and completely stop the post-harvest transition of sugar to starch, giving an extremely sweet crop.

The botany of saving seeds

It has been a long-held tradition to save the seeds from crops grown in the kitchen garden, and in many cases this is a simple and rewarding task. Seeds are produced after an ovule is fertilised, and an embryo is produced that contains the next generation of offspring.

When saving seeds, it is important to know whether the plant is self-pollinating or cross-pollinates. Self-pollinating plants always come true to type from seed as they only contain the genetic material of the parent plant. Those that cross-pollinate contain genetic information from two parents, giving variable offspring. In many cases, the pollen received in the ovary is still from the same cultivar, growing in close proximity, meaning the resulting seed and offspring have very similar properties to the parent plant.

But cross-pollinators can produce offspring that have marked differences, either when two cultivars cross, or in some instances, two different species cross. In some cases, these crosses can produce fantastic new cultivars, but in others, the qualities of the offspring can be disappointing, with poor colour or taste, and in some cases, even being poisonous.

Cross-pollinators can produce offspring that have marked differences, either when two cultivars cross, or in some instances, two different species cross.

There are some, very rare cases of plants such as squash, *Cucurbita pepo*, which have been bred to remove their natural toxin, cucurbitacin, which have cross-pollinated with a wild squash, and the resulting offspring have been bitter and poisonous. This is infrequent, and the most likely outcome of cross-pollination is that the young plant differs slightly from its parent. To overcome cross-pollination, the desired crop can be grown in isolation, meaning the only pollen that will reach it is from the same cultivar, rather than in the field where the pollen could have come from miles around.

Cross-pollination is also used in the horticultural industry to create F1 seeds and plants. These crosses are carefully controlled by the breeder to produce uniform young, which often show hybrid vigour. For example, a small purple carrot bred with a long orange carrot may produce a carrot that is longer and more purple than

Once harvested, seeds must be dried as much as possible, to avoid rot, and kept in cool, dark surroundings.

the parents. Seeds from F1 plants, known as F2s, do not come true to type.

When saving seed, it is important to ensure the seed is as ripe as possible to have a fully developed embryo with the largest food stores possible. The fruit that contains the seed often reaches horticultural ripeness, which is when it is at the best stage for eating. But seeds should not be harvested until they attain physiological ripeness, often when the fruit is brown, as in legumes. Once harvested, seeds must be dried as much as possible, to avoid rot, and kept in cool, dark surroundings. This reduces respiration to a minimum and allows the seeds to avoid using up their food stores and hence last for as long as possible.

Glossary

Achene, a classification of dry fruit that contains a single seed and does not split open upon maturity
Allelopathic, a plant that exudes a chemical t hat stops other plants growing nearby
Anemophilous, a wind-pollinated plant
Annual, a plant that completes its life cycle in one year or season
Apical dominance, where the central stem dominates the side growth
Auxin, a plant hormone that controls many aspects of growth including cell division
Axillary bud, or lateral bud, is a bud that grows in the axil, the place where the leaf or other organ originates from the stem; it can produce stems or flowers
Biennial, a plant that completes its life cycle over two years
Bracts, a modified leaf in the inflorescence
Bulb, modified leaves that sit upon a flattened stem growing underground and use as a storage organ
Calyx, the outer whorl of a flower, often subdivided into sepals
Cambium, the layer of tissue that produces new cells
Capitulum, the flower type of the Asteraceae family in which lots of small flowers sit on top of a flattened stem, giving the impression of one large flower
Carpel, the female reproductive unit consisting of ovary, stigma and usually a style
Chlorophyll, a green pigment in plants that absorbs sunlight for photosynthesis
Chloroplast, where chlorophyll resides in the plant cells and the site where starch is produced during photosynthesis
Cladode, a modified stem that is flattened and acts as a leaf, including photosynthesising
Collenchyma, a type of plant cell with several functions, including support
Corm, an underground storage unit formed from a swollen stem
Corolla, the petals that form the second whorl of the flower
Corymb, an inflorescence formed of several smaller flowers on pedicels of differing lengths but where all flowers are borne at the same height
Cotyledon, a seed leaf
Cultivar, in taxonomy, a plant bred in cultivation; *Genus*, *species*, 'Cultivar'
Cytokinin, a plant hormone with many functions, including cell division
Dehiscent, fruits that split open to release their seeds when ripe
Dicot, having two seed leaves
Dioecious, the male and female reproductive organs are borne on separate plants
Drupe, a classification of fruit that is fleshy and indehiscent, with the seed contained in a stony endocarp; also often called stone fruits
Endocarp, inner layer of a fruit wall
Entomophilous, pollinated by insects
Ephemeral, a plant that completes its life cycle extremely quickly, usually within one season or year
Epidermis, the outer layer of cells on a plant
Epigeal, referring to germination, when the seed leaves appear above ground

Etiolation, plant growth in the absence of light causing longer internodes and no green colouration
Exocarp, also known as the epicarp, the outer layer of cells of a fruit wall
Fruit, the part of the plant that contains the seeds
Genus, to do with taxonomy, a group of related species; *Genus*, *species*, 'Cultivar'
Gibberellin, a plant hormone involved in many developmental stages, including flowering
Halophytic, a plant adapted to live in a salty atmosphere such as the coast
Hermaphrodite, in flowers, a complete or bisexual flower that has both stamens and pistils
Hypanthium, also known as the receptacle, the base of the flower that contains all the flower parts
Hypocotyl, in a seedling, the part of the stem below the seed leaves or cotyledon
Hypogeal, referring to germination, where the seed leaves remain below the surface
Indehiscent, relating to fruits, those that do not split open when ripe
Inflorescence, the arrangement of the flowers emerging from one axis.
Internode, the area of the stem between two nodes
Involucre, also known as the phyllary, a group of tightly packed bracts that surround a compact flower head
Lamina, the leaf blade or edge
Legume, a type of dehiscent fruit that is formed from a single carpel and generally splits open along one side
Lenticel, corky areas on the surface of a plant, generally woody ones, where gaseous exchange occurs
Liana, a woody climber
Lignin, a polymer found in plant cells that creates woody growth
Locule, the space in the carpel where the ovule sits
Meristem, tissue in a plant that is undifferentiated but can divide
Mesocarp, in fruit, the middle layer of a fruit wall, often fleshy
Mesophytic, plants adapted to conditions that are neither too wet nor too dry
Monocot, plants with a single seed leaf, and sharing features such as parallel leaf veins and flower parts that come in multiples of three
Monoecious, both male and female plants on the same plant; the individual flowers may be hermaphrodite or separate males and females
Node, the place on the stem where the leaf is, or once was, attached
Nut, a classification of fruit, being indehiscent, with one seed and a dry pericarp
Ovary, the part of the female fruit that bears the ovules, and will eventually become the fruit
Panicle, an inflorescence that has a central axis with branches originating and then forking again before bearing a flower
Parenchyma, plant cells that have thin walls and make up the softer areas such as leaves
Parthenocarpy, to form a fruit without fertilisation
Peduncle, flower stalk, also known as the pedicel
Perennial, a plant that lives for many years
Perianth, the calyx and corolla
Pericarp, the wall of a developed fruit, consisting of the epicarp, mesocarp and endocarp
Perisperm, occurring only in some seeds, when the nucleus develops to become food store tissues and lies outside of the endosperm
Petiole, a leaf stalk
Phloem, part of the vascular system of a plant, that transports nutrients
Photosynthesis, the process through which plants convert water, sunlight and carbon dioxide into oxygen and energy in the form of sugars
Pyrene, the stone of a fruit that consists of the seed and a hardened layer of endocarp

Raceme, an inflorescence arrangement where the flowers form on pedicels coming from a central axis, and the flowers mature from the bottom up
Receptacle, the end of the flower stalk where the organs sit
Respiration, the breakdown of stored sugars using oxygen to release energy, water and carbon dioxide
Rhizome, an underground stem
Schizocarp, a fruit type that is formed of two or more carpels and splits into individual sections when mature
Sclereid cell, a reduced sclerenchyma cell, which is pitted and can give a gritty texture
Sclerenchyma, plant cells that are lignified
Sepal, a part of the outer whorl of a flower, an individual part that makes up the calyx
Spathe, a large bract that forms a sheath around a flower or spadix
Spadix, a fleshy spike bearing minute flowers
Species, in plant taxonomy, plants within a genus that are distinctive but can often interbreed; *Genus*, *species*, 'Cultivar'
Stomata, pores in the leaf that allow the movement of gases and water
Subspecies, in taxonomy, a category below species, where the plants fall within the same species but are distinct from one another due to developing in either different geographical or ecological conditions; *Genus*, *species*, subsp. *xxx*
Syconium, a fruit type in the family Moraceae where the inflorescence is enclosed and forms a multiple fruit
Synonym, in plant nomenclature, an old scientific name
Testa, the seed coat
Trichome, a hair that is formed from the epidermis and has no vascular tissue
Tuber, an underground storage organ formed by a modified stem
Variety, in taxonomy, ranks below subspecies, and consists of plants within the same species, that have differing characteristics, but none that make them distinct enough to be their own species; *Genus*, *species*, var. *xxx*
Vernalisation, a period of cold needed by plants to induce flowering in the following growing season
Xerophytic, a plant adapted to growing in areas with little available water
Xylem, part of the vascular system that primarily moves water from the roots to the leaves

Index

First published in 2023 by Royal Botanic Gardens, Kew, Richmond, Surrey, TW9 3AB, UK

www.kew.org

ISBN 978-1-84246-783-1

Distributed on behalf of the Royal Botanic Gardens, Kew in North America by the University of Chicago Press, 1427 East 60th Street, Chicago, IL 606037, USA.

British Library Cataloguing in Publication Data
A catalogue record for this book is available from the British Library.

Design: Ocky Murray
Production Manager: Georgina Hills
Proofreading: James Kingsland
Copy-editing: Matthew Seal
Printed in Italy by Printer Trento

For information or to purchase all Kew titles please visit **shop.kew.org/kewbooksonline** or email **publishing@kew.org**
Kew's mission is to understand and protect plant and fungi, for the wellbeing of people and the future of all life on Earth.

Kew receives approximately one third of its funding from Government through the Department for Environment, Food and Rural Affairs (Defra). All other funding needed to support Kew's vital work comes from members, foundations, donors and commercial activities, including book sales.

Picture credits

© Hélèna Dove: 17, 18, 20, 32, 34, 35, 37, 46, 49, 50, 54, 55, 57, 64, 66, 69, 72, 75, 76, 88, 89, 92, 96, 100, 101, 103, 108, 109, 113, 115, 117, 122, 124, 129, 131, 135, 136, 137, 154.

© The Board of Trustees of the Royal Botanic Gardens, Kew: 2, 4, 6, 8, 11, 13, 14, 24, 42, 60, 121, 125, 138, 150, 159.